MW00850069
23

THE MILITARY MACHINE, VOLUME ONE

U.S. Half-tracks

The development and deployment of the U.S. Army's half-track vehicles

Written by
David Doyle
Edited by Pat Stansell

Dedicated to the crewmen of U.S. half-tracks of all nationalities. Without your experiences and sacrifices, there would be no books such as this.

Published by The Ampersand Group, Inc.,
A HobbyLink Japan Company
235 NE 6th Ave., Suite B, Delray Beach FL 33483-5543
www.ampersandpubco.com
ISBN: 978-0-9895547-6-3

Front and back cover images: The robust construction of U.S. half-track vehicles has resulted in a great many survivors, including this superbly restored M2A1. It was photographed at the Museum of the American GI in 2009. (Author photo)

Title page: M3A1 half-tracks of an undisclosed unit plow through the snow near Born Belgium on 21 January, 1945. (NARA)

Front end paper: Partly completed M2 half-tracks proceed along the assembly line at the Diebold plant. Among other components, the fenders and radiator guards have yet to be installed. A skate rail appears to be already present in the interior of the lead vehicle. (LOC)

Rear end paper: A nearly complete half-track equipped with a winch has reached the end of the assembly line at Diebold. Note the early-production, civilian-type headlights. (LOC)

Table of contents: Two M2 half-tracks of Battery A of the 78th Field Artillery Battalion pause while tanks pass on a crossroads. This photo was taken during the September 1941 maneuvers in Louisiana. Both vehicles are pulling 75mm M1897 field guns. (Office of History, U.S. Army Corps of Engineers)

Image Credits: National Archives and Records Administration (NARA); Library of Congress, Office of War Information collection (LOC); Bundesarchiv (BA); American Truck Historical Society (ATHS); Patton Museum of Cavalry and Armor (Patton Museum); Rock Island Arsenal Museum; TACOM LCMC History Office; US Army Transportation Museum; US Army Ordnance Museum; (Ordnance Museum); US Army Quartermaster (Quartermaster); Reg Hodgson; John Adams-Graf; Jim Gilmore; Tom Wolboldt; Mike Peters; Diebold, Inc.; The Tank Museum; Image Bank WW2; Office of History, US Army Corps of Engineers; Heslop Collection, Brigham Young University; USMC; Imperial War Museum (IWM); Third Cavalry Museum; Clark County Historical Society; Wisconsin Historical Society; National Museum of the United States Air Force; Finnish Archives (SA-kuva-arkisto), Pat Stansell (PAS) and Tom Gannon.

Book design by Patrick Stansell. Line drawings by Todd Sturgell.

Table of Contents

Preface

Above: *Development of U.S. half-tracked vehicles went on for years prior to WWII. Shot on 30 April 1934, this photo depicts the T4 half-track truck. It was taken during wire-laying tests of the vehicle for the U.S. Government. (NARA)*

Opposite: *Admiral William F. Halsey, commander of the South Pacific Area, reviews U.S. troops from the rear of an M3A1 scout car at a base in the Southwestern Pacific in 1942 or 1943. Next to him is Col. Glenn Cunningham, U.S. Army. Note that the national star is within a circle of a slightly different shade than the color of the scout car. (NARA)*

The U.S. half-track remains one of the most iconic symbols of the Second World War. Although fielded by the French prior to 1940 and later by the Russians in very limited numbers, it was the U.S. Army and the German army that fully realized the armored infantry concept. The Germans were much more enamored with the half-track as a truck, creating a bewildering array of types, virtually all of which could be used to perform several tasks. The U.S. Army concentrated on just two basic types, although they were also used in a variety of roles. Like the Germans, one of these types was used specifically as an armored personnel carrier. In fact, the U.S. Army had little interest in the half-track as an unarmored truck, relying instead on its superb fleet of 4x4 and 6x6 tactical trucks.

U.S. military planners eventually designed the half-track fleet exclusively to be armored vehicles. They had many advantages over those used by the Germans. Perhaps most notable was that the front axle was powered. This is a huge advantage in a cross-country truck—especially one with tracks. The overall U.S. design philosophy was to improve the inherent capabilities of the 4x4 truck. This was accomplished without creating additional drive train components. The final representation used a conventional 4x4 transmission and axle arrangement, yet yielded a much more capable vehicle for overland travel.

U.S. half-track vehicles served on all fronts during WWII, with the U.S. Army, the USMC and many Lend-Lease partners. Although primarily used as a prime mover and an armored personnel carrier, the robustly designed chassis also was used as a platform for artillery, anti-tank guns and anti-aircraft weapons. Service continued briefly into the post-war period until eventually replaced by fully tracked vehicles. Other nations made use of the many surplus half-tracks, perhaps most significantly the Israelis, who re-engined their M3s and M3A1s and used them in the spearhead of their armored infantry units throughout the fifties and sixties.

There have been countless books on the subject of U.S. half-tracks over the last thirty years. Given the half-track's unique design and its extensive use, these armored trucks have been the subjects of intense interest with enthusiasts. In spite of this proliferation, much of this older material is now out of print. One of the purposes of this title is to redress that as thoroughly as possible. This is the first of two extensive volumes, both with the goal of covering the subject in as thorough manner as possible. Rather than have to rely on several incomplete publications, we have endeavored to provide the complete subject in these two volumes.

This first volume will be restricted to the presentation of the basic types and will omit any information on the gun motor carriages and other self-propelled artillery vehicles. This will be covered exclusively in the second book in the series.

The "gold standard" for U.S. vehicle reference has always been the excellent series authored by the late Richard P. Hunnicutt. Regrettably, the half-track title is also now out of print. Using that title as a benchmark, we have provided extensive coverage of all the half-tracks by type. This begins with the early development of the concept and we were fortunate to obtain many rare and interesting photographs of these unusual vehicles.

Our publication contains a "book within a book" on the development and deployment of the M3 scout car. The story of the scout car is closely tied to that of the half-track, as the original concept for the M2 half-track was based on a tracked version of the scout car. These two machines also shared the same manufacturer, White Motor, further tightening the family ties. The chapter found here is probably the most complete print coverage of the subject to date.

The commentary continues with a photo essay on the White factory, along with individual chapters on the M2 and M2A1, the M3 and M3A1, the M5 and finally the M9. The final chapter of the book briefly covers post-war use.

The biggest deviation from the Hunnicutt benchmark is our depiction of field use. In addition to the hundreds of development photos, we have added just as many combat and in-use shots. Drawing on a large variety of archival sources, we have endeavored to create as complete a picture as possible of the way these vehicles were used by the troops. We hope the reader will be especially pleased to see more than the typical coverage of the M5 and M9 series—so often ignored in previous publications. In addition to drawing on wartime photos, we have discovered many new pre-war photos of half-tracks during military exercises and war games.

Along with our extensive photographic reporting, we have relied upon a massive amount of archival documentary data to tell the story of the half-track. A great deal of useful information on half-track production and modification can be found within the appendices.

In order to provide the clearest picture possible we have incorporated as much information as we could into this publication. Inevitably, this meant selecting images that have appeared in print in the past. It would be irresponsible to not include the greatest breadth of material, so we hope the long-time collector will excuse seeing familiar images. One of the primary goals of these two volumes was to provide the reader with a single source for all of the relevant material on the subject.

In addition, thanks to extensive research, in many cases our images are based on the original negatives, which when combined with modern printing techniques, allow many of these "old favorites" to be reproduced with new clarity, and reveal previously hidden details.

Acknowledgements

It is not possible for a single person, or even a pair of authors, to write a book of this scope. Rather, what you hold now in your hands is the result of an extensive and broad collaboration. Were the input of any of the friends and colleagues listed below be removed, then this book would be measurably different.

Joe DeMarco, who has spent countless hours researching U.S. Army Ordnance during WWII, provided thoughtful insights into the production, registration and Ordnance serial numbers of the half-tracks.

Tom Gannon, author of the seminal two-volume *Israeli Half-tracks*, graciously dipped into his files and provided photos documenting arguably the final combat user of these vehicles.

Jim Gilmore, who has a vast collection of archival documents and photos, not only opened his collection for this project, but also took time out of his own research at the National Archives to dig up additional elusive documents.

Third Cavalry Museum Director Scott Hamric supplied previously unpublished photos of half-tracks both in combat and during stateside training.

Reg Hodgson, editor of *Army Motors*, a publication of the Military Vehicle Preservation Association, and noted vehicle collector, provided scarce photos and valuable insights on the Canadian use of the White Scout car.

Tom Kailbourn's careful and detailed analysis of many of the photos in this volume was invaluable, and he contributed immeasurably to the information presented herein.

Kevin Lockwood, prominent half-track historian, collector and preservationist, reviewed and corrected the manuscript, and provided thoughtful insights into the production quantities. This included, not insignificantly, proving that the vaunted 1944 *Armored and Tactical Vehicle Acceptances* data concerning M3 half-tracks is erroneous.

John McLeaf, who has spent years studying White scout cars, gladly shared his knowledge, considerably enhancing that portion of this book.

Eric Reinert allowed us to browse the files of the Corps of Engineers History collection, yielding previously unpublished material.

Rebekah Smith at Diebold, Inc., graciously provided scarce photos documenting that firm's significant contribution to the production of both scout cars and half-tracks.

Exceeding his role as just a publisher, Pat Stansell lent his skillful design talents, dug into his own extensive photo collection and offered valuable insight into numerous captions.

Scott Taylor lent a critical eye to the manuscript, providing corrections and amplification from the combined standpoints of writer, historian and modeler.

Fellow author, vehicle collector and historian Pat Ware supplied elusive documentation on the Commonwealth use of half-tracks, covering both the WWII and post-war eras.

Tom Wolboldt has spent countless hours analyzing U.S. military vehicle production during WWII and has developed extensive databases as a result of this research. Tom willingly shared this information, as well as items from his personal document and photo collection, to further this volume.

Lee Young, librarian and archivist for the American Truck Historical Society, graciously opened the Society's extensive White holdings for review, including numerous previously unpublished photos.

The former staff of the Patton Museum, Frank Jardim, Charles Lemons and Candace Fuller, allowed unfettered access to their extensive holdings, including the immense collections donated by Col. Robert Icks and historian R. P. Hunnicutt. When a personal visit was not possible, Don Moriarty graciously copied whatever materials were needed.

The staffs of the Military History Institute, Rock Island Arsenal Museum, Library of Congress, U.S. Army Ordnance Museum, TACOM LCMC History Office and the United States National Archives were without exception enthusiastic supporters of this effort, allowing access, providing suggestions and quickly responding to many queries.

However, the most enthusiastic and unflagging support provided for this project came from my wife Denise, who traveled coast-to-coast, endured hot, dusty archives, and willingly scanned thousands of documents and hundreds of photos during the six-year process that resulted in this volume. Without her support in this project, it would not have been possible.

—David Doyle
Memphis, Fall 2014

Chapter 1: Early Development

Concepts, Prototypes and Evaluations

Development and production of U.S. military half-tracks was relatively short-lived, spanning less than two decades. The bulk of the work and production occurred between 1939 and 1943. That some of these vehicles remained in service for several decades beyond makes the accomplishment even more remarkable. While U.S. manufacturers produced a bewildering array of half-tracks (or Half-Tracs in the terminology of White Motor Company) during WW2, all production models were based on chassis developed for personnel carriers or prime movers. This volume examines the development, manufacture and usage of these two types of half-track during World War II.

Half-tracks are combination wheeled and track-laying vehicles. Early models were propelled by the tractor track alone, but in later models propulsion was by the synchronized drive of the tractor track and front wheels.

The experience of World War I taught the combatants (and observers) that rapid reconnaissance and deployment of forces over rough terrain were critical components to victory. The fully tracked tank and other track-laying vehicles had made their debut, and they all had in common slow speeds and cumbersome steering. Military planners desired a vehicle that combined the off- road mobility of fully tracked vehicles with the maneuverability and on-road performance of a truck.

Initial efforts focused on "convertible" vehicles, typified by the efforts of J. Walter Christie. These vehicles rolled along roads on wheels and were fitted with tracks when off-road operation was anticipated. In the U.S., this system was largely advocated by the Cavalry Board and culminated in the April 1937 approval of the Car, Combat, Convertible, T7. Tests prior to the Plattsburg Maneuvers convinced Major General Adna Chaffee that convertible vehicles were not suitable, echoing a view espoused by Major (later Major General) Robert "Bob" Grow six years previously. Both men felt that the time and difficulty required for the wheel to track conversion were prohibitive obstacles. Renewed interest and considerable effort were then placed in the development of the half-track.

The Russian army first used military half-tracks in 1917. André-Gustave Citroën of France and Alexander Kégresse of Russia refined the concept, and in 1920 their collaboration resulted in an adaptation of the Citroën B2 automobile in a half-track configuration. In 1922, five Citroën-Kégresse vehicles successfully completed the Trans-Sahara Expedition crossing the desert to Timbuktu. Two years later, Georges-Marie Haardt and Louis Audoin-Dubreuil used Citroën-Kégresse half-track vehicles to traverse Africa in an eight-month journey that certainly validated the half-track concept.

It is not surprising then that in 1925 the U.S. Army Ordnance Department bought two of the Citroën-Kégresse vehicles for evaluation purposes. These 10-horsepower vehicles, driven through a rear drive pulley with torque transmitted by friction, were tested at Aberdeen Proving Ground as prime

Above: *Christie converted several Mack Model AC trucks to half-tracks in the early 1920s, including this example. It featured two solid wheels on each side of the cargo body with continuous tracks around them. (NARA)*

Below: *The U.S. Military tested Citroën-Kégresse half-tracks as early as 1925. This version, the P-17, was evaluated at Aberdeen in 1931. It has a 28-horsepower engine, not the 10-horsepower model used in earlier tests. (The Patton Museum)*

movers for the 75mm gun. The rear bogie was of the semi-rigid type and the front wheels were not driven. A further test article was purchased from Citroën in 1931. This half-track had a five-passenger body and boasted a more powerful 28-horsepower engine, while the pulley drive was moved forward of the bogie rollers.

In 1932, Ordnance Committee action 9957 of 7 July authorized the purchase and test of a domestic half-track produced by James Cunningham, Son and Company, Rochester, NY. Cunningham's initial offering was a 50-horsepower unit based on a Ford 1 ½-ton truck, but the vehicle eventually procured by Ordnance was propelled by a more powerful 115-hp 353 cu. in. Cadillac V-8. This vehicle, designated the T1, featured rubber block track, leaf spring suspension and sprung rear idler and was equipped with front sprocket drive. The T1 was followed by the T1E1, 30 examples of which were produced at Rock Island Arsenal using parts purchased from Cunningham. The T1E1 differed from the T1 in that the body had a canvas top and was designed to seat six people.

OCM (Ordnance Committee Minutes) 10482 of 10 February 1933 discussed the shortcomings of the track and idler. Accordingly, four of these vehicles were modified with new track and a rigid idler, resulting in the Half-track Car, T1E2. Further action, OCM 11207 of 12 June 1934, directed that volute spring suspension be fitted to one example as a test article, which was then designated T1E3. This suspension test would ultimately form the basis for the suspension system used on the WWII M2 and M3 half-track vehicles.

The vehicles discussed thus far have been half-track cars, but OCM 10630 of 20 April 1933 authorized Rock Island Arsenal to build a half-track truck. This vehicle combined a 1933 GMC 2 ½ T-33 truck with a half-track mechanism similar to that of the half-track car T1E2, with the resulting vehicle being designated Half-track Truck T1. The T1 was tested at Aberdeen from August through November 1933.

Also undergoing tests at Aberdeen in the closing months of 1933 was the Half-track Truck T2. The T2 was created by marrying a 1933 Ford 1 ½-ton truck with a Cunningham rear suspension like that used on the T1E1 half-track car. A similar combination was also tried in 1934 with a Chevrolet track as the basis, but the Ford adaptation proved superior.

Still in search of the ideal half-track truck, OCM 10742 of 25 May 1933 authorized the purchase of a model WD-12 commercial vehicle produced by the Linn Manufacturing Company of Morris, New York. A large vehicle featuring steel track and powered by an American LaFrance V12 engine. The vehicle, which the army designated T3, was capable of speeds of 15-20 MPH.

In late 1933, the Signal Corps had a requirement for a cross-country wire laying vehicle. To meet this need, the Half-track Truck T1 was redesigned and an aluminum body for wire laying was added. As recorded in OCM item 11160 of 15 December 1933, the resulting vehicle was designated T4.

The half-track truck T1 was redesigned in order to meet the needs of Field Artillery, as recorded in OCM 11589 dated 11 July 1934. In this effort, a GMC truck was modified to provide a "kick up" in the frame in order to raise the front drive sprocket. Then the leaf spring suspension as used on the Half-track Car T1E2 was installed, along with a special aluminum body. The resulting vehicle was designated "Truck, Half-track T5." Initially the Cunningham-designed rubber track blocks were installed, but those were later replaced with the Goodrich endless band tracks that would come to typify U.S. half-tracks during WWII.

A second vehicle was also procured for Field Artillery testing. Designated the Half-track Truck T6, this vehicle which was authorized by OCM item 11314 was an improved T3 with special field artillery body.

In 1938, the Half-track Truck, T8 underwent testing at Aberdeen Proving Ground. The T8 was a Ford truck with a Trackson rear drive unit and was determined to be unsatisfactory.

The next step was the testing of the T9, which introduced to the vehicle a feature that would become

Above: *The Citroën-Kégresse P17 half-track appears during testing at Aberdeen Proving Ground in June 1931. The grill had a distinctly nonmilitary design. (NARA)*

Below: *A close-up of the right rear side of a Citroën-Kégresse P17 half-track at Aberdeen in June 1931 shows the 9-inch-wide track, produced by Kégresse-Hinston. (NARA)*

75

a hallmark of U.S. half-tracks—the driven front axle. Authorized 4 December 1935 by OCM 12542, the Marmon-Herrington Company created the T9 by adding their front wheel drive unit to a Ford 1 ½-ton truck, while installing at the rear volute spring suspension and Goodrich endless band track as used previously on the T1E3 half-track car. The vehicle was also tested with a two-wheel bogie and rubber block, bushed track, a configuration known as the T9E1. Ultimately, the T9 and T9E1 proved both the possibility and the advantages of having a driven front axle on a half-track vehicle. OCM item 12801 of 22 April 1936 approved Specification AXS 429 and the procurement of additional half-track trucks T9.

In 1938, the first U.S. armored half-track was constructed. Per OCM 14188, personnel at Rock Island Arsenal in conjunction with White Motor Company fitted a rear bogie assembly of the type found on the half-track truck, T9 to a M2A1 scout car and modified the rear body accordingly. Significantly, the scout car's front wheel drive was retained. The reconfigured vehicle was designated the Half-track Personnel Carrier, T7. During tests at Aberdeen in September and October the type showed considerable promise, although it was noted that the vehicle was underpowered. During these tests an unditching roller, a feature later to be widely used on scout cars and half-tracks, was added. Although at the conclusion of testing the T7 was converted back into a M2A1 scout car, the die had been cast for further half-track development.

Above: *The P17 had a 3,500-pound drawbar pull, lending it to use as an artillery prime mover. Maximum speed was a modest 18 miles per hour. (The Patton Museum)*

Below: *The T1, manufactured by James Cunningham, Son and Company, was the first domestically produced military half-track. The vehicle was larger than the P-17 and had a more powerful Cadillac V-8 engine. (The Patton Museum)*

ORDNANCE DEPT. U.S. ARMY
HALF TRACK CAR T-1
US 67

Above: *The Cunningham T1 half-track car featured four dual bogie wheels on each side, center-guide rubber-block track assemblies, front sprockets, and sprung idlers mounted on long suspension arms perforated with lightening holes. (NARA)*

Below: *On the follow-up to the T1 half-track car, the Cunningham T1E1, the cab doors were eliminated. It featured stronger idler arms and a fuel tank and a stowage box on each rear fender. Thirty T1E1s were produced. (NARA)*

Above: *A T1E1 is displayed with the bows and tarpaulin erected over the body and a canvas top installed over the cab. A piece of steel channel served as the rear bumper; bolted to it was a step for accessing the body. (Rock Island Arsenal Museum)*

Below: *A single door provided access to the rear of the body of the T1E1. The bottom corners of the door were cut at angles. (Rock Island Arsenal Museum)*

Above: *The T1's successor, the T1E1, was produced by Rock Island Arsenal using components bought from the Cunningham company. The T1E1 has an enlarged rear body. (TACOM LCMC History Office)*

Below: *Rubber band tracks T14 made by Goodrich were experimentally installed on the T1E1 half-track truck. The durability of these tracks was unsatisfactory, and further development work commenced. (Ordnance Museum)*

Above: *A T1E1 half-track car is being put through its paces during tests or maneuvers. Crates are strapped to the forward end of the rear fender, and a unit marking, a square with the number six, is on the hood. (US Army Transportation Museum)*

Below: *In a photo taken during U.S. Cavalry tests, a T1E1 has been armed with a .30-caliber machine gun. The insignia of the 1st Cavalry Regiment, an eagle in an eight-pointed star, is on the cowl. (US Army Transportation Museum)*

Above: *A 1st Cavalry trooper demonstrates a firing position for the Browning .30-caliber machine gun on a pedestal bolted to the right side of the cowl of a T1E1 half-track car. The steel-mesh grille is clearly shown. (US Army Transportation Museum)*

Below: *Two of the T1E1 half-track cars evaluated by the 1st Cavalry Regiment are parked together. They are marked, right to left, number 27 and number 20. (US Army Transportation Museum)*

Above: *The T1E2, a modification of the T1, sports a rigidly mounted idler in lieu of the spring-loaded unit used on the base model. (TACOM LCMC History Office)*

Below: *The venerable T1 pilot was fitted with a volute spring articulated bogie, becoming the T1E3. This suspension is the basis for the suspension employed by production U.S. half-tracks during WWII. (TACOM LCMC History Office)*

Above: *This T1E3 is almost new in this photograph, which was taken at Rock Island Arsenal in August 1934. That this vehicle was based on the T1 is apparent due to the rear body configuration. (The Patton Museum)*

Below: *The T1 series of half-track cars are not the only half-tracks designated T1. Pictured here is the T1 half-track truck at Aberdeen Proving Ground in August 1933. (The Patton Museum)*

Above: *Powered by an American LaFrance V-12 gasoline engine, the Linn T3 was a large and imposing machine. It could attain a top speed of almost 20 miles per hour on level roads. (Ordnance Museum)*

Below: *As seen in a June 1933 Aberdeen Proving Ground photograph of the Linn T3 half-track truck, the tracks were fitted with rollers and were routed around two teardrop-shaped skids mounted on the bogie frame. (Ordnance Museum)*

Above: *This image is illustrative as to why the Linn T3 was a half-track truck, not a half-track car. The cargo here is the T1E6 light tank. (The Patton Museum)*

Below: *James A. Cunningham, Son & Co. of Rochester, New York, produced several half-track trucks, including this version based on a 1932 Ford 1 ½-ton 4x2 truck. It is shown at Aberdeen Proving Ground in August 1933. (Patton Museum)*

Above: *A December 1933 Aberdeen Proving Ground photo shows a left-hand Cunningham half-track mechanism, which featured a center-guide oil-bushing track assembly and detachable grousers. (NARA)*

Below: *Cunningham also produced this half-track truck based on a Chevrolet truck chassis. This example and the preceding one based on the Ford chassis had idlers mounted on long, spring-loaded suspension arms. (Patton Museum)*

Above: *The T4 half-track truck was an improvement of the T1 half-track truck, designed for better performance and longevity. Based on a General Motors chassis, this vehicle was photographed at Aberdeen Proving Ground in June 1934. (NARA) (Patton Museum)* **Below:** *The T4 half-track truck is viewed from the left side. It featured a General Motors 6-cylinder engine, T10 rubber-band track assemblies, and size 7:00-20 tires. (NARA)*

Above: *A T4 half-track truck is viewed close-up from the right side, showing the suspension in detail. The rails above the body supported the tarpaulin bows. The large loops on the side of the body were for lashing down the tarpaulin. (NARA)*

Below: *The right bogie assembly of a T4 half-track truck has been broken down for inspection in a 1934 Aberdeen Proving Ground photo, showing the wheels, spindles, arms, and springs. (NARA)*

Above: *The left track, bogie, idler, and sprocket of a T4 half-track truck are depicted. A good view is available of the side of the cargo body, with its heavy-duty loops for lashing the tarpaulin. (Jim Gilmore collection)*

Below: *A civilian driver is maneuvering a T4 half-track truck toward a steep railroad embankment. The photo was taken during wire-laying tests of the vehicle for the U.S. Government on 30 April 1934. (NARA)*

Above: *The T5 half-track truck was similar to the T4 except with several new features, including cab doors; an enlarged, 401 cubic inch engine; convertible-type frames for the cab top, and cowl-mounted headlight assemblies. (NARA)*

Below: *Another T5 half-track truck, U.S. Army registration number W-4084, is observed from the left front. The embossed GMC logo was painted at the center of the front bumper. Soft-plastic windows were in the tarpaulin and door curtains. (NARA)*

Above: *T5 half-track truck U.S.A. number W-4084 has the full canvas installed. A filler tarpaulin was attached to the rear of the cab top and to the front of the main tarpaulin over the body: a rather clunky, makeshift design.* **Below:** *As seen from the right rear in a March 1935 photo, the T5 half-track truck had upholstered lazy backs and troop seats; the seats are raised in this photo. The vehicle was equipped with a tailgate. Below the tailgate is a tow pintle. (NARA, both)*

Above: *Another T5 half-track truck, U.S. Army registration number W-4078, presents its front end. The GMC logo is on the upper shell of the radiator and on the front bumper. Details of the construction of the grille and the fenders are in view. (NARA)*

Below: *A photo of the rear of a T5 half-track truck shows close-up details of the tailgate and retainer chains, the single tail-light assembly, the tow pintle hook, the rear face of the chassis frame, and the mud flaps. (NARA)*

Above: *In a slightly elevated view of the rear of a T5 half-track truck, the upholstered seats and lazy backs, the stowed bows for the tarpaulin, and the diamond-tread floor are visible. (NARA)*

Below: *The bogie suspension of the T5 half-track truck was a near duplicate of the leaf spring suspension used on the T1E2 half-track car. (Jim Gilmore collection)*

Above: *The chassis of a T5 is observed from above with the hood, cab, and body removed. The fuel tank is to the front of the rear axle, below where the cab seats were located. The battery is on the right running board. (NARA)*

Below: *The fuel tank has been removed in this close-up view of the cab and engine areas of a T5 half-track truck chassis, enabling a view of the transmission and the drive shaft. (NARA)*

With the bodywork of a T5 half-track removed, one may see elements of the vehicle from the engine to the cab to the rear axle, along with the chassis frame and cross members and the fenders and running boards. (NARA)

Above: *In another close-up view of the transmission and rear axle of a T5 half-track truck, some of the linkages for the control levers are in view. (NARA)*

Below: *Twenty-four T5s were built, including this Battery B, 68th Field Artillery Battalion vehicle, shown here during the 3rd Army maneuvers in 1940. (The Patton Museum)*

Above: *During one of the U.S. Army maneuvers in the continental United States, probably Tennessee, a T5 half-track truck pulls a 75mm pack howitzer M1. Bed rolls are strapped to the side of the cargo body. (NARA)*

Below: *The same T5 half-track truck towing a 75mm pack howitzer passes several light tanks during maneuvers.. The T5 was designed as a light-artillery prime mover, and the 1,440-pound pack howitzer was well within its capacity. (NARA)*

Above: *The Linn T6 half-track truck is similar to the earlier T3, but slightly smaller. The vehicle performed admirably during testing at several Army facilities, but was hampered by a top speed of only 15 miles per hour top. (The Patton Museum)*

Below: *As was the case with the earlier Linn T3 half-track truck, the tracks, sprockets, idlers, and bogie wheels of the Linn T6 half-track truck were located inside of the chassis frame. (Jim Gilmore collection)*

Above: *In 1938 the army was still interested in "convertible" vehicles, such as this T7 Combat Car. These vehicles would utilize its tracks, as shown here, when operating off-road.* **Below:** *When operating on-road, the Convertible Combat Car would ride directly on the road wheels, as here. However, the time and complexity required for the conversion led this type of vehicle to fall out of favor. (Rock Island Arsenal Museum, both)*

Above: *In an effort to bring together the best of both worlds, technicians from White Motor Company and Rock Island Arsenal joined forces to convert a M2A1 scout car to half-track configuration. The similarity to the wartime half-tracks is immediately evident.* **Below:** *The converted vehicle was designated the Personnel Carrier, Half Track T7, and tipped the scales at 12,170 pounds. The vehicle is notable for having driven front wheels. (The Patton Museum, both)*

Above: *The front wheel drive of the T7 Personnel Carrier was slightly preceded by the Marmon-Herrington T9 half-track truck. This vehicle is based on a 1936 Ford truck chassis. Following testing, the vehicle was standardized as the half-track truck M2. (Ordnance Museum)* **Below:** *T9 W-401611, with an M1917 75mm field gun hitched to it, is depicted in what is assumed to be its original configuration, before conversion to the T9E1, with two large bogie-wheel assemblies per side and solid, non-perforated, sprockets. (NARA)*

Above: *Tests on the washboard course at Aberdeen Proving Ground in January 1937 disclosed a tendency of the bogie wheels to jam while negotiating rough terrain, as shown here. (Jim Gilmore collection)*

Below: *The T9E1 half-track truck replaced the T9's four small bogie-wheel assemblies on each side with two large ones. On T9E1s, the front bogie wheel was closer to the sprocket than on the pilot T9, yielding a shorter track length. (NARA)*

Above: *T9E1 W-40161 is seen from the left side. These vehicles were equipped with front rollers to assist them in negotiating ditches. A substantial brush guard was installed to the front of the radiator assembly.* **Below:** *The Ford trademark is embossed in raised letters on the rear sill of a T9E1 half-track truck. Rather than a tailgate, two racks were installed on the rear of the body, with hooks to lock them together and to the side racks. (NARA, both)*

Above: *The T8 half-track truck as tested at Aberdeen Proving Ground comprised a modified Ford 1 ½-ton truck chassis with a flat cargo bed and a Trackson convertible wheel-drive or track-drive assembly. (Jim Gilmore collection)*

Below: *The Trackson track on the T8 half-track truck was 15 3/4 inches wide. Regarding the T8, Army records noted "Tests proved this truck to be worthless as a half-track vehicle." (Jim Gilmore collection)*

Chapter 2
The White Scout Car

Development and Production

The debate of the merits of wheeled versus track-laying armor has raged since the early days of mechanized combat, and even continues today. Just prior to the U.S. involvement in WWII, the development of armored scout cars was a part of this debate.

The lineage of the White M3A1 scout car can be traced directly to the Scout Car T7, tested from July through November 1934. The T7 was built by the Indiana Motors Corporation subsidiary of White. In a departure from previous scout car designs, which were built on passenger car chassis, the T7 was based on the Indiana model 12x4 1½ ton 4x4 truck chassis. Seventy-six T7 scout cars were produced and were standardized as the M1 in 1934.

A scout car built by the Marmon-Herrington Company was also tested from September 13 through September 30, 1937. This vehicle was designated the T13, and despite the difference in builder, it more closely resembled the subsequent M3A1 than did the Indiana T7. Testing showed that the T13's 221 cubic inch Ford V-8 was not an adequate powerplant for the 7,710-pound vehicle.

About a year later, the Corbitt Company's scout car, the T9, was tested, and was found to be superior to the T7. The Corbitt T9 was later redesignated as the M2, and 20 were produced.

The next major step in the development of scout cars was the White M2A1. While similar to the M2, the M2A1 incorporated a continuous skate rail, or tourelle, around the upper portion of the armored body. Upon this rail was installed the T38 gun mount, which included the D32985 pintle and 50 round .50-caliber ammunition box. The total weight suspended from the skate rail with this installation was 190 pounds. An adapter was furnished to allow the mounting of a .30-caliber M1919A4 in lieu of the M2 .50-caliber.

While a similar skate rail had been unsuccessfully tested earlier on the T13, the installation in the M2A1 proved successful, albeit the range of depression available, from -14 to +65 degrees elevation, was felt lacking for antiaircraft defense.

The pilot M2A1 was tested at Aberdeen Proving Ground from September 9 to September 30, 1937 under program 5353. It was determined that this vehicle was superior to any scout car previously evaluated by the U.S. Army. The vehicle was subsequently redesignated the "Scout Car M3." Thirty-six of the machines were produced in fiscal year 1937, 39 in fiscal year 1938 and 25 in fiscal year 1939.

A 95 horsepower Hercules JXD 6-cylinder engine powered the M3. The M3 was very similar to its successor, differing most obviously in the shape of the rear body. Whereas the familiar M3A1 has slab sides and wheel wells beneath the internal toolboxes, the M3 had a narrower body, with rectangular external wheel wells. At the front the M3 did not have the front mounted unditching roller, but at the rear, it did have a door. White built only 100 of the M3, but the basic shape of the scout car had been set.

Above: *The M1, pictured here, is the direct ancestor of the M3A1 scout car. Originally classified as the T7, the M1 was built by the Indiana Motor Corp, a subsidiary of White. Note the external machine gun pedestal mounts. (PAS)*

Below: *The M2, built by Corbitt, was the next vehicle considered. Its original designation was the T9. Although externally very similar to the T7, it was mechanically very different and had superior performance. (Patton Museum)*

W-60208

U.S.A. W-60278

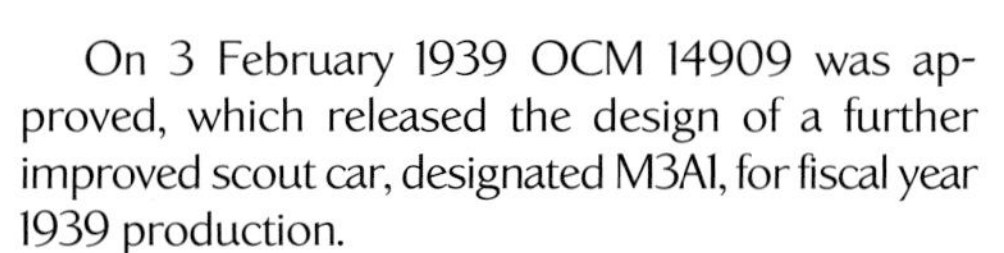

On 3 February 1939 OCM 14909 was approved, which released the design of a further improved scout car, designated M3A1, for fiscal year 1939 production.

The pilot model of the M3A1 was driven from the White plant in Cleveland, Ohio, to Aberdeen Proving Ground, Maryland for testing in mid-June 1939. Aberdeen Proving Ground personnel tested this vehicle between June 28 and July 18 under program 5395. The earliest M3A1s differed from later production models in various details. Initially, the "greenhouse" type windshield was retained. This system featured a protruding windshield with a side "wing" on each side. The windshield was designed to be folded forward when the armor cover needed to be lowered. Large wing nuts were used to secure the glass panel in the upper position, or two small "ears" on the side of the hood when it was to be secured in the lowered position. Later production vehicles had the more familiar removable glass panels that were to be removed prior to lowering the armor cover.

Also characteristic of the earliest vehicles was a right-side mounted spotlight and a lack of brush guards for the headlamps. Early M3A1s also had a cast shield mounted on the doors bearing the crossed sabers of the cavalry. Some references state that the using arm added these, but in reality they were installed during the manufacture of the vehicle, as supported by the accompanying photographs.

Internally, the compression ratio of the JXD was raised from 5.78:1 to 5.88:1. This, coupled with a new manifold for improved breathing, increasing the maximum RPM by 200, and an improved valve train raised the horsepower rating of the 320 cubic inch displacement Hercules to 110. All the M3A1 scout cars were 4x4, but unlike most four-wheel drive tactical vehicles, White used a full time four-wheel drive system. The operator could not disconnect power from the front axle. Also, the tourelle mount of the M3A1 made a rear door impractical.

On 27 July 1939 OCM 15235 was approved. This action covered minor modifications in the design that were to be incorporated in the initial production of the vehicles. Series production of the M3A1 scout car began in 1939, following receipt of a Telegram from Rock Island Arsenal notifying the company of an award on 24 August 1939. The initial production order was for 274 vehicles. That contract was quickly followed up with another order for an additional 25 vehicles. These 299 vehicles, plus the pilot vehicle, were all completed by April 18, 1940. Ultimately, a total of 16 contracts were issued to the White Motor Company in Cleveland, Ohio between August 1939 and January 1944.

White assembled the chassis, including front fenders and operator's controls. Once the chassis had been completed they were then driven to the nearby Diebold Safe and Lock Company. Diebold fabricated and installed the armor plating, which was ¼-inch thick and the face-hardened type. After installation of the armor, the cars were driven back to the White plant for final assembly and inspection.

In late 1942, scout car production fell short of plans by 120 vehicles due to a shortage of armor plate. A report by the Adjutant General's office stated in this regard "Shortage was due to production difficulties experienced at Diebold Safe & Lock Co., who install armor plate for White. Expect shortages to be made up in November and December 1942. Difficulties attributable to Diebold's failure to schedule material and production in accordance with White's schedule."

This was not the first time that the M3A1 scout car program had run into production difficulties. A report to the Adjutant General in September 1942, when 650 vehicles had been scheduled but only 295 built, had noted "The proposed program on Scout Car production will be difficult to fulfill for the reason that:

The project carries an A-1-c rating
Steel requirements an A-1-a rating
Brass, Copper and Bronze an A-1-c rating.

In order to accomplish the proposed program it will be necessary to provide an A-1-a rating for the entire project."

Not surprisingly, not all scout cars were alike.

Above: *Among vehicles of the 1st Cavalry Regiment assembled for an exercise in the 1930s are, far left, an M2A1 scout car, and next to it, M1 scout car U.S.A. number 60205. (PAS)*

Below: *The crewmen of an M1 scout car pose as if going about their duties. In the front are a lieutenant and the driver, and in the rear are a gunner and a radio operator. (Patton Museum)*

Ho
U.S.A. W 40201
50205

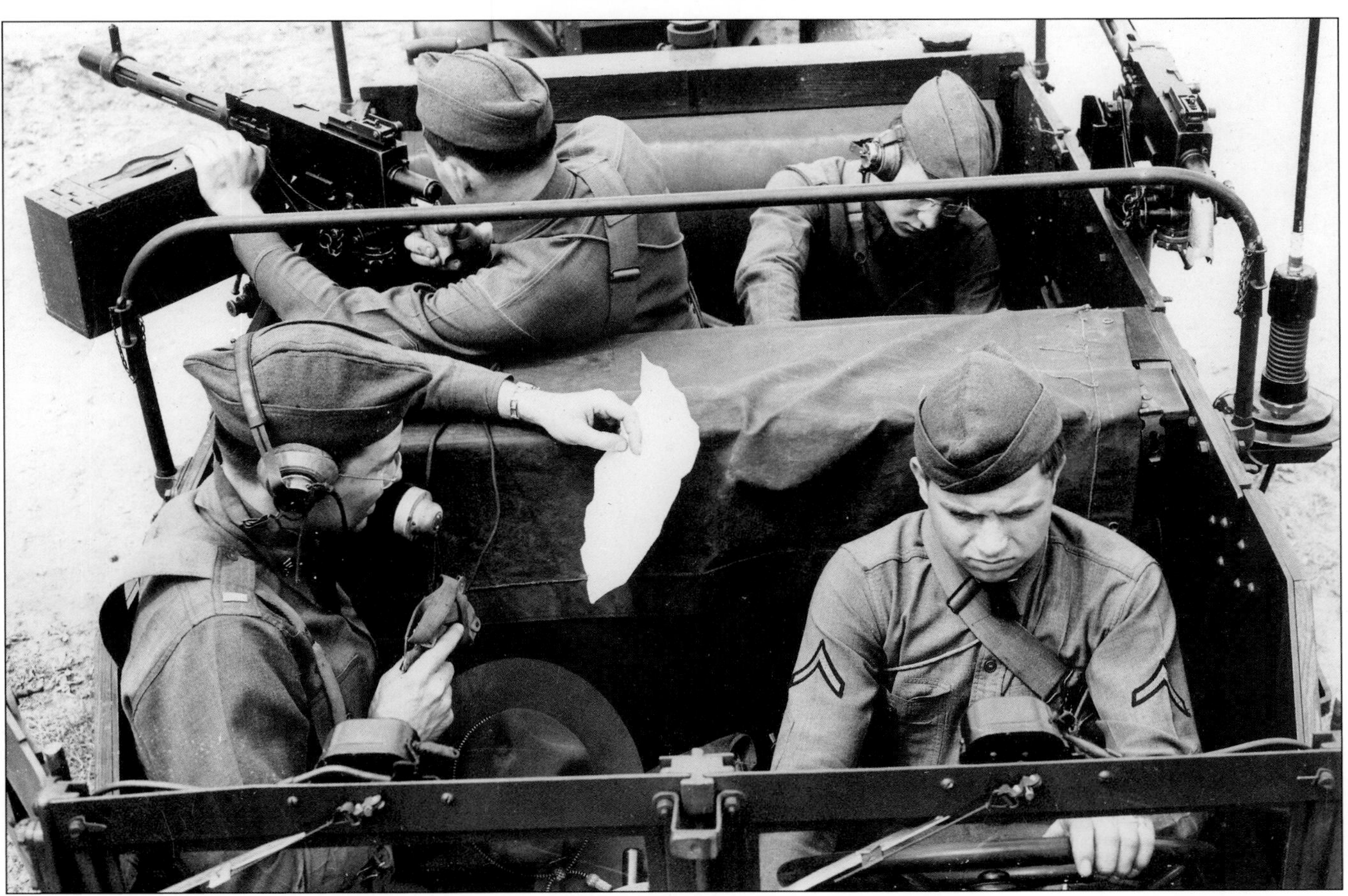

The ORD 9 SNL G-67 calls out 8 different production lots as detailed in the appendix of this book. The "Lot 1" vehicles, 298 scout cars on two contracts, did not have black out marker lights, and utilized civilian-style tail lights. The Lot 1 vehicles, as well as a portion of the Lot 2 vehicles, utilized Budd wheels with eight approximately triangular cut outs around their perimeter. The remainder of the Lot 2 vehicles used wheels with six oval cutouts. The wheels of both Lot 1 and Lot 2 are referred to as split ring type. Beginning with lot 3, the scout car was equipped with combat wheels, with the familiar heavy ring held in place by 18 bolts.

By the time M3A1 scout car production ceased in April 1944, 21,175 of the vehicles had been completed. Slightly more than half of these vehicles, 10,975, were supplied to other nations. The recipients were:

British Empire	6,997
USSR	3,340
French Forces	287
China	139
Brazil	84
Chile	50
Columbia	2
Costa Rica	2
Cuba	4
Dominican Republic	4
Ecuador	12
Guatemala	4
Haiti	2
Honduras	3
Mexico	9
Peru	30
Venezuela	6

Armament was a variety of machine guns mounted by means of trolley and pintle to a skate rail that went completely around the interior of the fighting compartment. Today, this skate rail has often been removed or had the section above the doors cut out. Another frequent civilian modification is cutting a door in the rear armor. All too often the entire rear armor has been cut away, reducing the scout car to a flat bed truck.

The M3A1 was a relatively fast vehicle, but was lacking in off-road performance when compared to tracked vehicles, and its open top left its crew vulnerable. By 1944, M3A1s were being sold as surplus.

Diesel-powered scout cars

In 1939 Ordnance became very interested in Diesel engines for combat vehicles. The reasons for this were summed up in OCM Item 15403, which stated "Test of Diesel engines to date indicate the desirability of employing this type of power plant more generally in the automotive vehicles supplied by the Ordnance Department. Tests of the General Motors commercial 2-cycle engine indicate that is generally satisfactory and a very reliable unit. The principle advantages of this type of engine are: Increased radius of action with a given fuel tank capacity; greatly reduced fire hazard; torque characteristics much superior to those of comparable gasoline engines; freedom from radio interference of the high tension ignition systems used on gasoline engines."

This OCM item further provided authorization to test Diesel engines in a variety of vehicles. Among those was the White scout car. On 3 November 1939 it was noted in OCM 15471 that "The Buda Company has agreed to supply, without cost to the Government, one of their 6DT-317 110 H.P. Diesel engines, for test in a Scout Car M3A1. Arrangements have been made to transfer the second car off the production line overland from the White Company Plant in Cleveland to the Buda Plant in Harvey, Illinois, for installation of the engine. Following installation of the engine and shop test of the modified vehicle, the car will be driven to Aberdeen Proving Ground by Proving Ground personnel for a technical test." The minutes continued "In order to keep the record clear, it is considered advisable to assign a designation to this vehicle." Accordingly, on 9 November 1939 the M3A1 powered by a Buda 6DT-317 was designated the M3A1E1.

Action by the Ordnance Committee on 9 November 1939 also authorized the installation of

Above: *The Marmon-Herrington T13, shown here, differs considerably in appearance from the other vehicles (note, for example, the elaborate fenders). Unlike the T7 and T9, which mount their weapons in stationary locations, the T13 scout car has a tourelle (skate) rail. (Patton Museum)*

Below: *White's M2A1 incorporated many improvements based on lessons learned from the previous designs. The vehicle was tested at Aberdeen Proving Ground in September 1937 and was found to be a substantial improvement over the earlier vehicles. The armor of the M2A1, as well as the M3 and M3A1, was fabricated and installed by the Diebold Safe and Lock Company. The M2A1's narrow (as compared to the familiar M3A1) rear body compartment is visible in this publicity photo taken in front of Diebold's office. (Diebold, Inc.)*

W 60552
IOWA
N.G.

CO.
CO.

a second type of Diesel engine in the scout car. This engine was the Hercules DJXD. Both Diesel-powered scout cars were tested at Aberdeen Proving Ground from January 1940 through April 1940, and found to be operationally equal to the gasoline powered version. In April 1940, the two Diesels along with a gasoline-powered machine were driven about 700 miles to Fort Knox, Kentucky for further testing.

A report of 3 July 1940 from the Sub-Committee on Automotive Equipment to the Ordnance Committee, Technical Staff included the following remarks with regard to the Hercules and Buda-powered scout cars: "The commercial Diesel engines....have been installed in Scout Cars M3A1, tested at Aberdeen Proving Ground and were used during the 3rd Army Maneuvers. The performance of these engines throughout these tests were satisfactory and the Chief of Cavalry now desires that this type of engine be used in the Scout Cars procured during the next fiscal year. The tests so far conducted have been limited to 2 Diesel-engined Scout Cars. It is the opinion of the Sub-committee that before deciding to equip all Scout Cars with Diesel engines that this type of engine should be subjected to a thorough service test. It is believed that this can be accomplished by equipping approximately 100 Scout Cars M3A1 with suitable automotive type Diesel engines which have been tested by the Ordnance Department and found satisfactory. Diesel-engined Scout Cars should be provided with axles of greater capacity to insure freedom from axle trouble due to the greater weight and torque load imposed by the Diesel engine."

Because of these factors, the Sub-Committee recommended that 100 M3A1 vehicles be equipped with Diesel engines "as early as practicable" and that the type be designated M3A2. This recommendation was approved by OCM 15948 on 11 July 1940. However, the M3A2 designation did not remain in place very long with regard to this vehicle. On 10 February 1942 Brigadier General Gladeon M. Barnes directed that the Diesel-powered vehicles be redesignated M3A1E5. With regard to OCM 15948, Barnes wrote "Inasmuch as the action recommended in the above reference covered only the design of a properly engineered Diesel engine scout car, under which authority one pilot is being manufactured, the designation Scout Car M3A2 was inadvertently and incorrectly assigned. The designation assigned this design should indicate an item of experimental nature rather than that of an adopted type.

"In view of the above, the designation Scout Car M3A2 is cancelled and the design of properly engineered Diesel engine scout car initiated by OCM Item 15948, is redesignated 'Scout Car M3A1E5.'

"Production Order C-1617, and Procurement Authority ORD 15153 P5-13 A 1005-01, relating to the procurement of a pilot vehicle, and any related papers or tabulations will be amended to indicate Scout Car M3A1E5 instead of Scout Car M3A2."

Barnes's direction was implemented on 14 February 1942 by OCM action 17813. However, interest in Diesel-powered vehicles had begun to wane, and on 12 March 1942 OCM 17919 recorded the Adjutant General's policy that all U.S. Army wheeled vehicles were to be powered by gasoline engines. On 14 April 1942 the Services of Supply asked that the development project for scout car M3A1E5 be cancelled.

Accordingly, on 20 April 1944 the development of the M3A1E5 scout car was cancelled, with the OCM action that accomplished this noting: "The development of Scout Car M3A1E5 (previously designated M3A2) was undertaken with the view toward providing a properly engineered Diesel engine Scout Car. This development contemplated the installation of a Hercules "DWXD" Diesel engine in a Scout Car M3A1 redesigned to accommodate the heavier and more powerful engine. The development and supply of the engine has been delayed due to the more urgent priority of other Government orders held by the Hercules Company."

Other scout car variants.

Perhaps the scout car variant most commonly illustrated in publications is the M3A1E2. This

Above: *One of the contributing factors to the success of the White scout car design was its chassis. Shown here is the chassis of the M2, which is based on a design by White's Truck division. Note the position of the battery near the frame on the driver's side of the vehicle. (ATHS)*

Below: *This photo clearly shows the protruding "greenhouse" type windshield that was used on the M2A1, M3, and early M3A1 vehicles. When in the normal position, as seen here, the glass supports the armored flap in the raised position. The windshield wipers are mounted at the top of the windshield, rather than the low mounting of later scout cars. (Patton Museum)*

vehicle had an armored hood installed over the open body, to protect against small arms fire and well-placed grenades. The sides of the cover turned down to present a smaller target, but were hinged and could be folded upward to increase the field of fire of the scout cars own weapons. Development of this vehicle was initiated on 3 October 1940 by OCM 16161, assigned the designation M3A1E2, and directed that this work be carried out at Aberdeen Proving Ground. Initial testing utilizing mock up armored roof configurations revealed that these components interfered with operation of the vehicles weapons, and adversely effected handling of the vehicle. Hence, on 22 December 1941 OCM 17611 cancelled the project, noting: "Several wooden and sheet metal mockups of a roof for the Scout Car M3A1 have been made at Aberdeen Proving Ground... however no actual armor or soft plate construction was undertaken, due largely to the priority of other development items."

An M3A1E3 was also built. This vehicle was armed with a 37mm cannon mounted in the rear compartment on a pedestal, and was intended to give the Cavalry a mobile, high-speed vehicle with antitank capabilities. Development was initiated by OCM 15865 on 6 June 1940, and recommended the M3 37mm gun. This recommendation was changed to include the 37mm gun mount T6 by OCM 16261 of 14 November 1940, the same action also assigning the M3A1E3 designation. Testing of the vehicle was undertaken at Aberdeen in mid-1941, and included installation of a standardized 37mm gun mount M25. Ultimately, however, it was determined that the resulting vehicle offered too high of a silhouette and the weight of the weapon and ammunition hampered vehicle performance.

Above: *Unlike its successor, the M2 features a rear door. However, due to the position of the skate rail and the rear seat back (which can fold forward), entry and exit from the rear was awkward. The scout car is furnished with a tarpaulin to cover the crew compartment during bad weather while in rear areas. The M2 and M2A1 also have a rear door, as can be seen here. This feature is not present on the M3A1. (ATHS)*

Below: *Taken during evaluation of the various machine gun mounts, this M3 photo provides a good perspective of the right side profile. The M3 added brush guards for the headlamps and eliminated the skate rail's weight-reducing holes. Note the bracket just forward of the door for the collapsible canvas water container. (NARA)*

U.S.A.W-60511

Right: *This M3 is on trial at Ft. Knox, Kentucky. This angle shows the cramped quarters inside the vehicle that resulted from the external rear fenders. The armored windshield is closed in this view. (Patton Museum)*

Below: *The armored flaps that protected the radiator, were controlled by a lever inside the cab, are in the closed position in this photo. Crewmen were cautioned against extended operation with the louvers closed due to fears about the engine overheating. Note that the headlight guards on the M3 are mounted vertically. (Patton Museum)*

Above: *Barely visible on the door of this M3 is the famous crossed-saber plaque. By this time, the vehicle's armament of a M2 .50-caliber machine gun and two 30-caliber machine guns had been established. Note the toolbox mounted on the driver's side running board. (Patton Museum)*

Below: *Members of the 16th Engineer Battalion, 1st Armored Brigade, 1st Armored Division, return a scout car to the opposite bank of the Ohio River after a mock night attack. This photo was taken during winter maneuvers at Ft. Knox in December of 1940. The car is the early M3. (NARA)*

Above: *The M3A1 is the first derivative of the M3. The vehicle in this photo is the pilot model delivered to Aberdeen Proving Ground for testing. It incorporates some features not found in most production vehicles, including a "greenhouse" type windshield and dual spotlights. However, the unditching roller has been installed, the headlamp brush guards have acquired their backward rake, and the rear body has been widened (the wheel wells are now internal). (NARA)*

Right: *Some sources state the spotlight was mounted on only one side of the M3A1. However, as can be seen in this head-on view of the M3A1 on the slope test at Aberdeen Proving Ground, the spotlights are in fact mounted on both sides of the cowl. (Patton Museum)*

Above: *Another view of the same scene. According to test reports, the M3A1 performed well in all its incline-based performance tests. Different forms of this vehicle, registration number W-60653, are shown in multiple photos in this chapter. (Patton Museum)*

Below: *This pilot M3A1 was driven from the White facility to Aberdeen Proving Ground for testing—note the Ohio manufacturer's license. Also note the stowage present on the rear for both the machine gun tripods and a cradle; the mounts for the second cradle are visible on the left side. Subsequent testing at Ft. Knox resulted in the recommendation that the stowed cradle be eliminated. Visible lower on the body are the civilian-type taillights and the braces for the rear bumper, which form steps for climbing onto the rear of the vehicle. Finally, note the rearward exiting exhaust pipe. (NARA)*

Above: *The production M3A1 was powered by the Hercules JXD 6-cylinder gasoline engine. The same basic engine, with minor variations, also powered the Studebaker US6 and the Ford M8 and M20 armored cars. (ATHS)*

Below: *The scout car's 320-cubic inch displacement JXD developed 86 horsepower at 2,800 RPM, and a respectable 200 foot-pounds of torque at 1,150 RPM. (ATHS)*

Above: *Workers lower the assembly containing the firewall and dashboard onto an M3A1 chassis. The radio filter and the crew heater are mounted below the dashboard. (LOC)*

Left: *Two assembly-line workers make adjustments to components inside the engine compartment of an M3A1 scout car that is nearing completion. (LOC)*

Left: *Workers at White's Cleveland factory lower a scout car powerplant onto a waiting chassis. The 6-cylinder engine, coupled to a four-speed transmission and two-speed transfer case, could push the vehicle to a rated 50 miles per hour. (ATHS)*

Below: *These M3A1 scout cars were photographed at a Diebold facility. The armor was not installed on an assembly line, but rather was mounted individually on each vehicle. (Diebold, Inc.)*

The front end of this M3A1 scout car was hoisted up to allow the underside of the chassis to be photographed, showing elements from the roller to the axles, suspension, and drive train. (Reg Hodgson collection)

Above: *These M3A1 scout cars were photographed at a Diebold facility. The armor was not installed on an assembly line, but rather was mounted individually on each vehicle. The inset shows an enlargement of the door of the second car in the photo. The crossed-saber plate is clearly installed as part of the armor assembly. Rock Island Arsenal records state that the plates were manufactured there and shipped to Diebold for installation. Although the wheels and rollers appear to be painted black, a comparison with the completed scout car in the background suggests that these components were painted pre-war gloss olive drab. (Diebold, Inc.)*

Below: *These in-progress scout cars are later, second contract vehicles, as revealed by their six-slot wheels and flat olive drab paint. Also visible are the factory-installed Willard batteries. The next stop for these trucks is the Diebold facility for armor installation. The exhaust pipe now terminates out the side of the vehicle, just forward of the passenger side rear wheel. (ATHS)*

Above *:Inspectors check over M3A1 scout car chassis. From the wheels, we know that these are Lot 2 vehicles, and the exhaust exiting out of the side is indicative of a vehicle produced after chassis number 227401. The change was to prevent exhaust from entering vehicle when the top was up. (ATHS)*

Below: *This vehicle awaits transit to the Diebold factory, which was accomplished simply by driving the incomplete scout cars (and later, half-tracks) across town. The tread pattern of the 8.25-20 Seibring Special Service tires is visible, as are the round horns mounted low on the firewall. (ATHS)*

Above: *Recently completed M3A1 scout cars with tarpaulins and cab tops installed are lined up. U.S. Army registration numbers have been applied to the hoods in blue-drab paint. At chassis serial number 227401 the outside rear view mirrors and spotlights were deleted, and dull olive drab paint was introduced, as recorded in a service bulletin issued 28 June 1940. (NARA)* **Below:** *Once complete, the scout cars were driven back to the White plant where they underwent final inspection. Any deficiencies were corrected prior to government acceptance. (ATHS)*

Newly minted M3A1 U.S.A. number W-602081 is the subject of a patriotic display at the Cleveland Trust Company building at the corner of East Ninth and Euclid in Cleveland, Ohio. (NARA)

Above: *A broadside view of another early scout car, complete with chrome-plated spotlights and light-colored wheel locking rims on the wheels. The sheen on the front fenders is indicative of the gloss paint. All these traits would change when the United States went to war. (ATHS)* **Below:** *Another view of the pilot model M3A1. Note the open-sided construction of the tarpaulin, which has separate side curtains and an isinglass rear window. The tarpaulin was extensively revised prior to series production. Beneath the passenger door is an axe, and below that is a hatchet. The hatchet was eliminated by recommendation of the Mechanized Cavalry Board and APG personnel. (NARA)*

Above: *This is an early production M3A1. While it retains the civilian-type taillights and the spotlights, the greenhouse windshield assembly has been eliminated. Note the "eight-hole" wheels, which are characteristic of the first two orders for this type vehicle. The exhaust pipe is not visible, indicating this vehicle has a rear exhaust. (ATHS)* **Below:** *The canvas tarpaulin has been installed on this early scout car. The tarpaulin protected the crew and radio equipment from the elements, but it also drastically reduced visibility. (ATHS)*

Above: *A production model truck. The brackets for the hatchet have been eliminated, and the tarpaulin is completely redesigned. Formerly, two tarpaulin support bows were used, but production vehicles have three. The side curtains now roll up, rather than detach as they did before. Despite appearances, this truck does have a registration number. It is applied on the side of the hood in blue drab, which is almost invisible in black and white photography. The last two digits, 98, are faintly apparent. (NARA)* **Below:** *An early production M3A1. It retains the civilian-type taillights, but the rear stowage has been standardized. Still present are the spotlights and the "eight-hole" wheels. Note the aforementioned reconfigured rear bumper bracing. The exhaust pipe is not visible, indicating this vehicle has a rear exhaust. (NARA)*

Above: *This M3A1 is seen patrolling an airfield in the Panama Canal Zone in 1940. The U.S. took the security of the canal very seriously and increased its presence there as soon as hostilities started in 1939. The shroud of the .30-caliber weapon is the very early slotted type. (NARA)* **Below:** *This vehicle displays the rims that, according to ORD 9 SNL G-067, are fitted to all M3A1s from the third contract forward. These are commonly referred to as "combat rims." Regardless of style, most rims are dated. The liquid container (jerry can) brackets were introduced per White Service Bulletin 28, on 11 November 1942. (Patton Museum)*

Above: *An overhead view of the same vehicle reveals the outward-facing crew seats just behind the driver's and commander's seats. Also visible are the storage lockers above the fender wells and the first aid kit mounted on the outer wall, just behind the driver's seat. (Patton Museum)*

Below: *By comparing this photo of a M3A1 chassis to the earlier photo of the M2A1 chassis, many differences become apparent. Among these are the relocation of the vehicle battery to the passenger's side, and the dual center-mounted fuel tanks. (ATHS)*

Above: *The Dutch government ordered a number of M3A1s, including this example. In addition to two water-cooled Browning .30-caliber machine guns, it had a Lewis machine gun, barely visible on the right side of the skate rail. Also present were small lights inboard of the headlights. (Reg Hodgson collection)*

Below: *This M3A1E2 underwent testing in May 1941. Its armored roof offered protection to its occupants from splinters and small arms fire. The narrow openings still provided a small vulnerable area. Reportedly, this vehicle survives (without its roof). (Patton Museum)*

Above: *The side "flaps" of the M3A1E2's armored roof were hinged, and could swing upward to provide a wider field of fire for the .50-caliber M2 up front and the pair of .30-caliber M1919A4 machine guns on the sides. (Patton Museum)*

Below: *This photo, taken in June 1941, shows vehicle W-60653. It has now been converted to the M3A1E3 configuration by the installation of the 37mm anti-tank gun mount T6. The vehicle still retains its chrome spotlights and prewar white registration number. (NARA)*

Above: *The date was long ago cropped out of this photo of W-60653, but by this time the M3A1E3 has been repainted in flat olive drab, including the chrome spotlights. The registration number has been reapplied in blue drab. Note the contrast in color between the blue drab registration number and the white type number. (Patton Museum)* **Below:** *The mounting for the 37mm anti-tank gun is visible in this overhead view of the E3. There has been significant revision to the rear compartment seating. (Patton Museum)*

Above: *Originally designated the Scout Car M3A2, this vehicle, U.S.A. number 609226, was the pilot for a Diesel-engine-powered version of the M3A1. However, the designation was changed to M3A1E5 in February 1942. (Reg Hodgson collection)* **Below:** *While there is a strong resemblance between M3A1 scout car and half-track, the front armor is in fact not interchangeable.*

This photo of an M3A2 appears to reveal the use of half-track components on this vehicle, including grill and headlights. The vehicle also appears to utilize a half-track front axle. Development of the M3A1E5 ceased on 20 April 1942. (TACOM LCMC History Office)

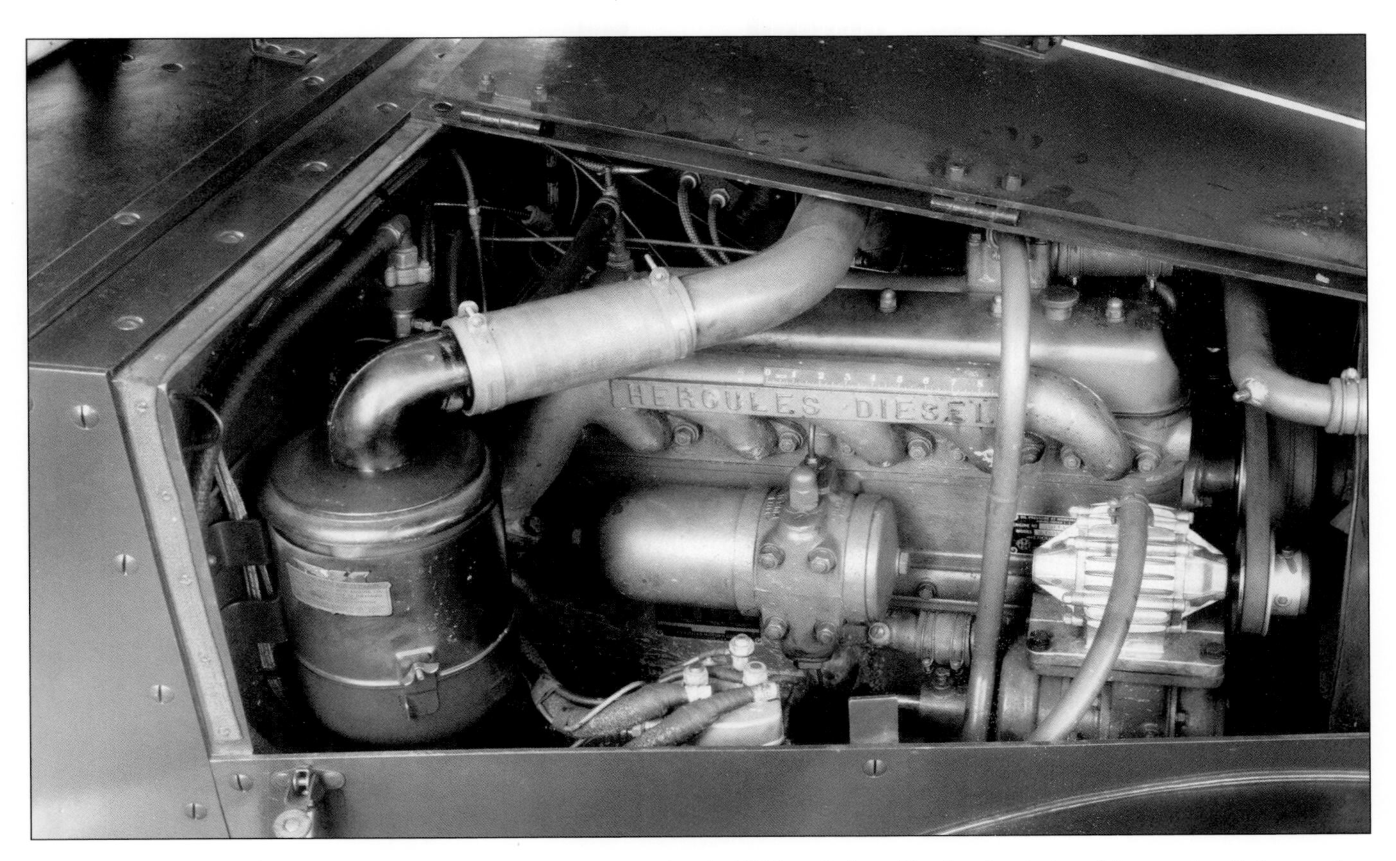

Above: *The Hercules DJXD, the powerplant used in the M3A2, is seen here from the passenger's side of the vehicle. At least one of these vehicles survives in chassis form in the hands of a private collector. (NARA)*

Below: *The driver's side view of the same installation reveals the Bosch fuel injection pump. One hundred of these DJXD-powered scout cars were reportedly built, but none saw combat with U.S. forces. (NARA)*

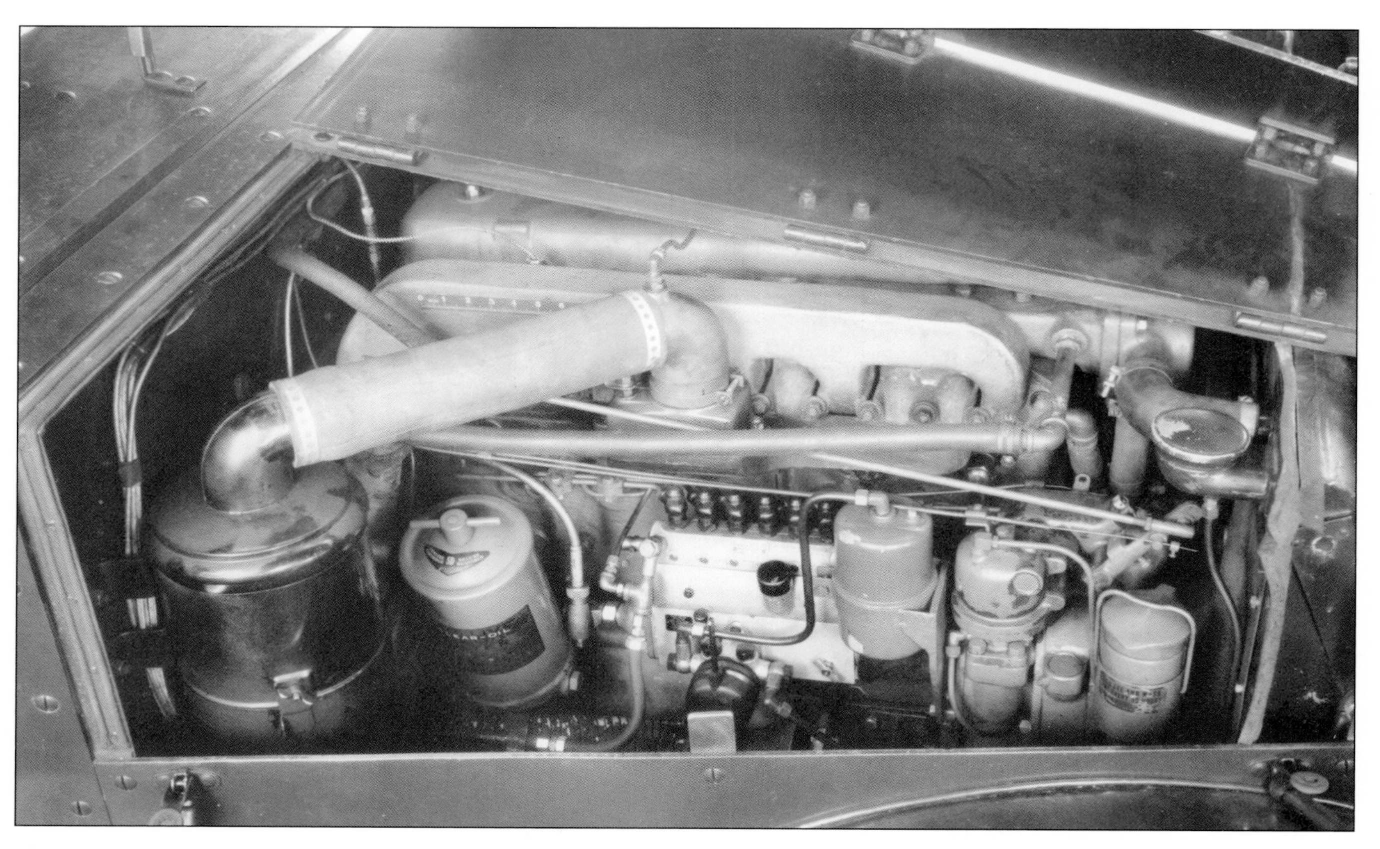

Above: *The diesel-powered M3A1E1 has a Buda-Lanova 6DT-317 engine. The fuel injection pump is on the right side of the engine. This photograph was taken in April 1940. (NARA)*

Below: *The left side of the Buda-Lanova 6DT-317 diesel engine installation in the M3A1E1 scout car is shown. Visible is the belt-driven generator (bottom) and fuel booster pump. (NARA)*

The Buda-powered vehicles were not produced beyond the pilot model. Although Diesel engines were successful in the scout cars, diesel was not at this time the standard fuel of the U.S. Army. (ATHS)

Left: *This scout car is being used to evaluate a deep water wading kit. What appears in many photos to be rivets holding the scout car armor together are in fact slotted, oval head screws. The screws are all oriented in the same direction, with the slots running vertically. Two theories for this have been advanced. The first states that it lessens their tendency to retain rain-water, and the second that it makes it easier to ascertain any loosening of the screws. (NARA)*

Below: *A large assembly of U.S. Army vehicles in the field includes a number of M3A1 scout cars in the foreground. More M3A1s are present at various points in the background, along with various types of command cars and trucks. (NARA)*

Above: *A crew poses in an M3A1 assigned to headquarters, 1st Armored Regiment (Light), as the 1st Cavalry was known after 15 July 1940. Photographed during the Louisiana maneuvers in October 1940, the vehicle sports eight slot wheels, chrome spotlights, rearview mirrors, and cavalry placards on the doors in addition to the painted crossed sabers. The insignia of the 1st Cavalry, a hawk inside an eight-pointed star, is on the cowl. (NARA)* **Below:** *An M3A1 scout car with an Air Corps-style recognition star on the shutters (white star with a red circle in the middle, over a blue circle) crosses an engineer bridge outside of Camp Polk, LA in September of 1941. (NARA)*

Above: *This M3A1 is hitching a ride across one of the many rivers near Camp Polk, LA in the summer of 1941. The insignia on the door of the scout car indicates that it belongs to the 1st Cavalry Division. (NARA)* **Below:** *Troopers of Company F of the 102nd Cavalry move forward during the Louisiana maneuvers of August 1941. The regiment was designated as the 102nd Cavalry on 17 August 1921 from the 1st New Jersey Cavalry Regiment and had its headquarters in Newark. The regiment was initially assigned to the 21st Cavalry Division and was re-designated 102nd Cavalry Regiment (Horse and Mechanized) on 16 November 1940. (NARA)*

Right: *Major General John Millikin, Commanding General of the 2nd Cavalry Division, uses the radio of his M3A1. This photo was taken near Carthage, Texas on 27 September 1941. The Shreveport Times main headline reads "PLANES BLAST BLUE ATTACK," a reference to the massive war game going on in the area at the time. (NARA)*

Below: *A view of an adjacent vehicle shows the scout car's civilian-type instruments and its machine gun and antenna mounting. Of particular interest is the personnel heater. It is visible beneath the dashboard, directly under the .50-caliber machine gun. Strapped to the skate rail are pads intended to reduce discomfort during operation on rough terrain. In practice, these were also very helpful when entering or exiting the vehicle. (NARA)*

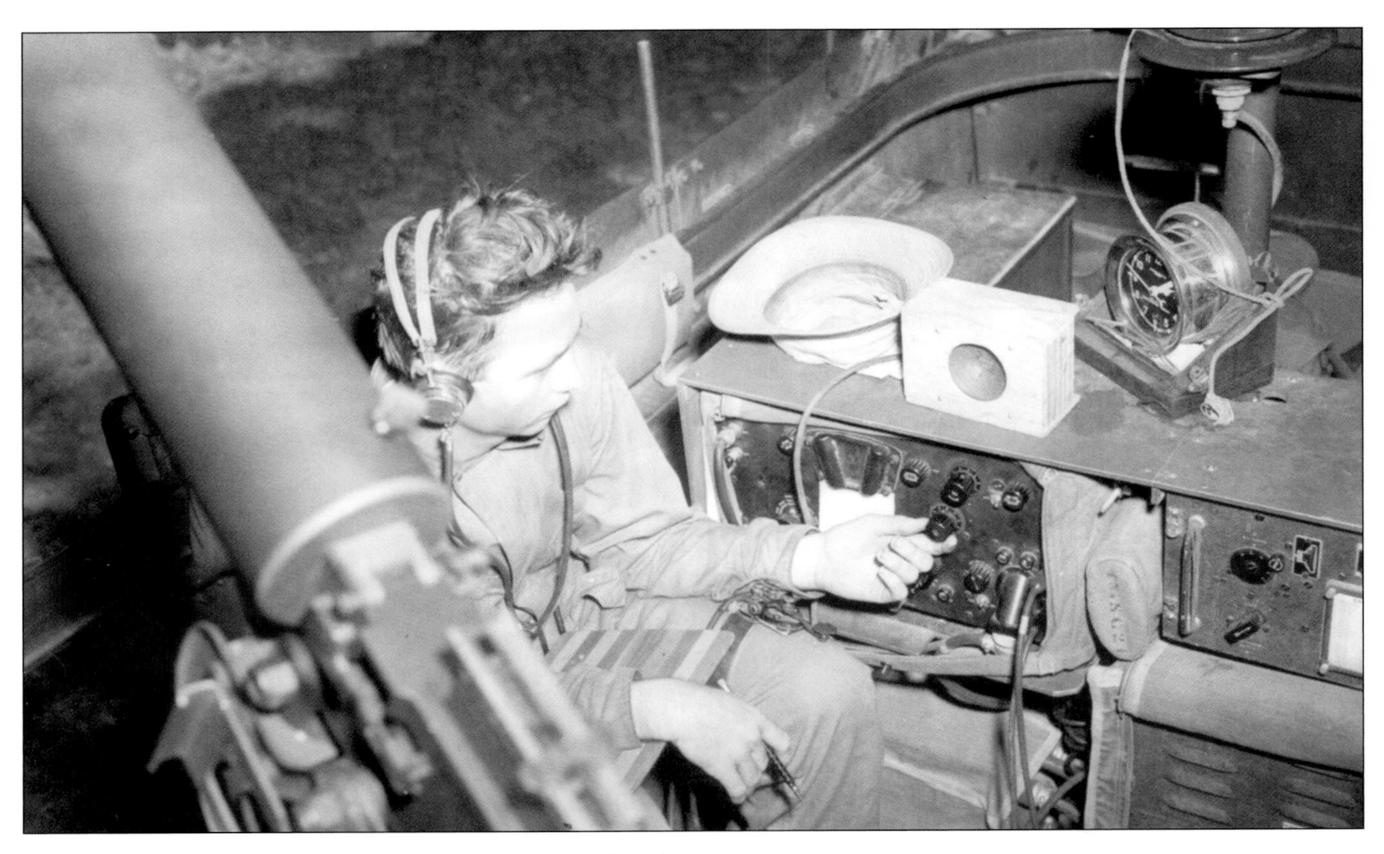

Above: *Here we see the scout car as viewed from the front looking back. The radio operator is receiving a Morse code transmission. Note the clipboard on his lap and the straight key device strapped to his leg. The U.S. Army message center clock and the wooden-housed speaker are non-standard accessories.* (NARA) **Below:** *Among many other units, the 2nd Cavalry Regiment participated with the Second Armored Division in the Louisiana maneuvers of September 1941. In this photo, horses with the 2nd Cavalry are emerging from the woods among new M3A1 scout cars.* (NARA)

Above: *The old ways of cavalry warfare meet the new during maneuvers as mounted troops ride past an M3A1 scout car with a motorcycle lashed to the rear armor. (NARA)* **Below:** *Officers view a map hung from the side of an M3A1 scout car of the headquarters troop of the 107th Cavalry. Originally raised in Ohio, the 107th Cavalry was inducted into federal service on 5 March 1941. This photo was taken during the Louisiana maneuvers in either August or September of 1941. (NARA)*

Above: *Scout cars and their M3 37mm anti tank guns of the 6th Cavalry Regiment sit on the grounds of Ft. Ogelthorpe, Georgia in November 1940. The 6th was stationed there from 1919 to 1942. Ft. Oglethorpe is only a few miles from the Tennessee border and was a staging area for many of the maneuvers held there. When deployed during WWII, the unit became part of George S. Patton's Third Army. (NARA)*

Below: *Major-General Charles L. Scott stands in an M3A1 scout of the car of the 2nd Armored Division during the Tennessee maneuvers of June 1941. General Scott had commanded the 2nd AD until the previous November, when Brigadier General George S. Patton replaced him. At that time General Scott became commander of the I Armored Corps, which counted the 2nd AD as a subordinate unit. (NARA)*

Above: *At various times, General George S. Patton used an M3A1 scout car as his personal staff vehicle. He would occasionally modify them to suit his personal tastes. Note the two non-standard horns mounted on the hood. During maneuvers at Manchester, Tennessee, on 19 June 1941, Patton (left) was photographed on an M3A1 bearing Army Air Corps-style national insignia. (NARA)*

Below: *This first or early second contract M3A1 is participating in the Tennessee maneuvers of June 1941. Like the previously shown White scout cars, it has the early-type, 8-slot wheels, but it lacks the cowl-mounted spotlights. The dark patch on the door could indicate that the crossed-sabers have been removed from this vehicle. The car commander is wearing a gas mask, presumably for protection from road dust. (NARA)*

Above: *G.I.s man two of the three machine guns of an M3A1 during exercises in Tennessee. The .30-caliber machine gun is an air-cooled model of the M1919 family. A stowage rack fashioned from steel strips is attached to the right rear of the body. Secured to the left fender is an M3 tripod mount for the .50-caliber machine gun, for use when the machine gun was dismounted from the vehicle and placed into action on the ground. (NARA)* **Below:** *During the Tennessee maneuvers in June 1941, an M3A1 scout car of the 2nd Armored Division fords a stream after finding that a bridge at that location had been "destroyed." A second M3A1 prepares to cross. The front bumper markings of the lead car read "2 DIV" and "HQ 20." The armored windshield covers and upper door panels of both vehicles are secured shut. (NARA)*

Above: *The next vehicle to cross is followed by several Ford GP jeeps, which appear to be lashed to the car by heavy rope. This scout car is marked with "HQ 21" on its left bumper. The star and roundel marking on the grilles can be seen more clearly here. (NARA)*

Below: *A scout car and its crew of non-coms strike a pose at an undisclosed location, most likely Tennessee. Although the date on the original caption is 1942, the presence of the M1917A1 helmets may indicate a date prior to the adoption of the M1 "steel pot" in June 1941. (NARA)*

Above: *This M3A1 was still in service in the fall of 1943 as the 749th Tank Battalion participated in the Tennessee maneuvers held from September to November of that year. While training and when deployed to Europe, the 749th was assigned to the 79th Infantry Division. (NARA)*

Below: *An M3A1 scout car defends the command post of the 15th Reconnaissance Company on 23 October 1941 during maneuvers. The location is South Carolina, but no specific town is mentioned in the original caption, as is sometimes the case. (NARA)*

Above: *Six scout cars pose in a sandy area. No markings are visible on any of the scout cars, but several of the soldiers are wearing what appears to be the 1st Armored Division patch. If they are members of that unit, this picture was probably taken during the South Carolina maneuvers during the fall of 1941.*

Below: *Gas masks get a test during 1st Army maneuvers in Camden, South Carolina. This scout car of the 102nd Mechanized Cavalry is among the vehicles participating there in early October 1941. (NARA, both)*

Above: *A scout car rolls off a rail car in Rock Hill, South Carolina in early November of 1941. Based on the design of the wheels and the absence of plaques, this is probably a second series scout car. Just visible near the headlamps are the early type, blue-lens blackout marker lights. The metal to the rear of the roller has been folded back. This redesign was a result of tests that showed that the original design forced water into the radiator during fording operations. (NARA)*

Below: *During II Corps exercises at Wadesboro, North Carolina, on 3 November 1941, a column of M3A1 scout cars has just driven over a mock minefield on the approach to an "enemy" position. In the ditch in the foreground, perilously close to the road, are defending troops with a .30-caliber machine gun. Markings on the front M3A1 include a white circle on the left bumper and E-104-11 on the right bumper. (NARA)*

Troopers conduct maintenance on their scout cars in the Carolina twilight. Although the pre-war maneuvers revealed serious shortcomings in the U.S. armored force, the experience of subsisting independently outside the garrison proved invaluable for the troops in the months to come. (NARA)

Above: *Members of the Red force defend a sunken road from within their scout car during the Carolina maneuvers. A strip of red cloth can be seen wrapped around the vehicle. A command pennant is also mounted above the driver's compartment. (NARA)* **Below:** *This trooper is making a bed of out of the hood of his scout car. The residual warmth of the engine is probably a benefit on those cold Carolina fall nights. The various platoons within the cavalry troops were assigned playing card symbols for identification. The accumulation of dirt and dust around the recesses and screw heads is of interest. (NARA)*

Above: *The closest vehicle in this column of M3A1 scout cars apparently photographed during Ft. Benning maneuvers is U.S.A. number W-60971; the number is painted white, which was the practice up to late 1940. (NARA)*

Below: *A column of 6th Cavalry vehicles, including an M3A1 in the lead, passes a column of troops of the 11th Infantry during Third Army maneuvers at Fort Benning, Georgia. The lead M3A1 bears the registration number W-60666, and spotlights with chrome housings are on both sides of the cowl. (NARA)*

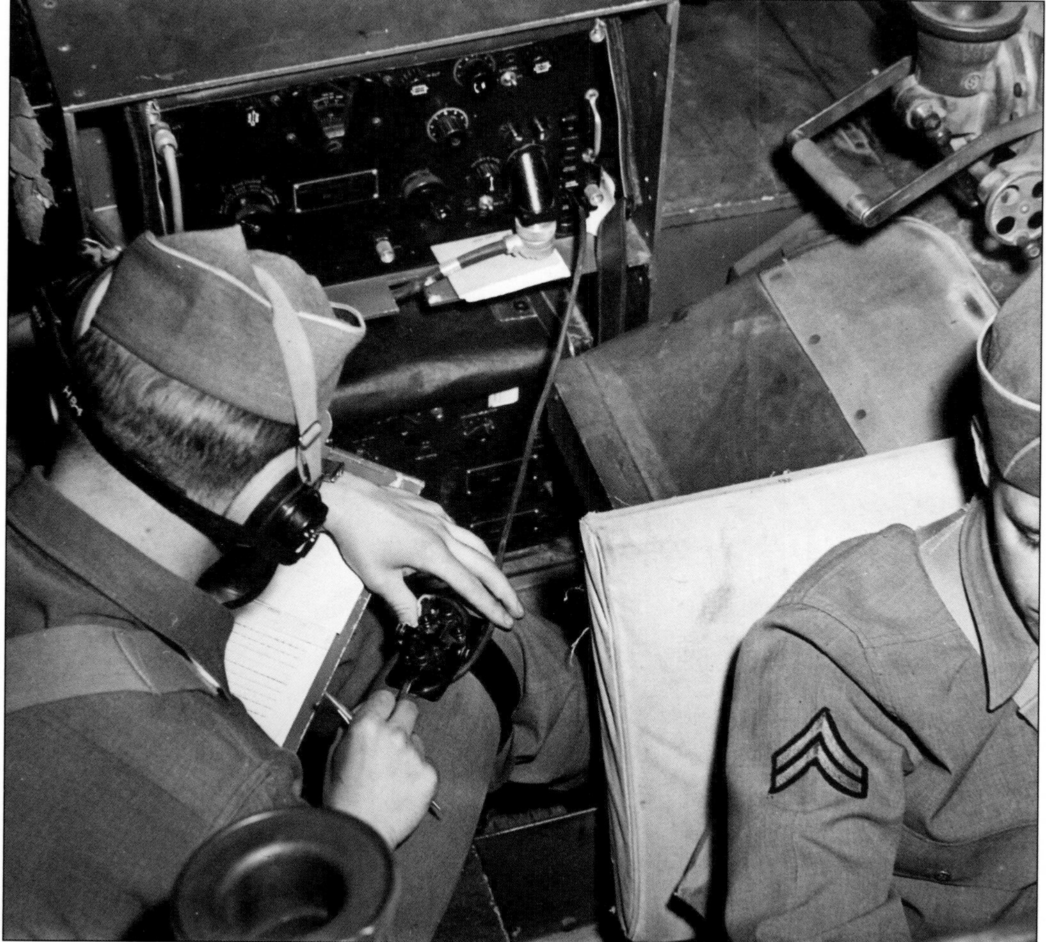

Above: *In this close-up of the skate rail above the driver's door on a Signal Corps M3A1 at Fort Riley, Kansas, the straps that hold the crash pad are visible. The crash pad protected the driver's head when it inevitably banged against the skate rail. (LOC)*

Left: *Another view of the previous scout car. The crew is part of the 2nd Cavalry Division, which was forming at Ft. Riley at the time. This photo provides a good close-up of the SCR-506 radio set. The operator is using the straight key device to send a Morse code message. Note the knee strap for the device and its lead. (LOC)*

Above: *An M3A1 scout car of a mechanized cavalry reconnaissance unit proceeds down a hill during a field problem at Fort Riley. Following it are a jeep and several motorcycles. (LOC)*
Below: *Working its way up a slope at Fort Riley, this M3A1 offers a clear view of its roller and radiator shutters. These shutters were typically left open except when combat conditions were anticipated. (LOC)*

Above: *Behind the windshield of an M3A1, a gunner aims the M1917A1 .30-caliber machine gun while another crewman is ready to observe his fire. (LOC)*

Below: *A crewman on an M3A1 of the mechanized cavalry is at the ready on an M1917A1 machine gun with leaf sight raised. Connected to the cooling jacket of the machine gun is a rubber hose, called the steam condensing device, the lower end of which was connected to an M1 water chest. (LOC)*

Above: *Another view of the preceeding photo.The groups of four slotted, oval-headed screws spaced at intervals along the upper part of the body serve to attach the skate rail to the body. (LOC)*

Right: *A radioman transmits a message from an M3A1 scout car during a sham battle at Fort Riley, Kansas, in April 1942. The transmitter and receiver units are housed in a steel rack near the rear of the crew compartment. Partially visible to the right of the radio set is a M1917A1 machine gun tripod mount. (LOC)*

A crewman hands the same radio operator a message for transmission. Note the ammunition locker on the right side of the compartment and the radio antenna and floor mast mount to the front of the radio set. (LOC)

Above: *The gunner of an M3A1 scout car takes aim during a sham battle at Fort Riley. Details of the skate rail and the vision shutter and bracket on the inside of the upper door panel are visible. The name "Jane" is painted below the tactical symbol of a downward-pointing arrow. (LOC)*

Below: *At Fort Riley, Kansas, during the April 1942 war games, an M3A1 scout car is proceeding virtually sideways while negotiating a muddy road. Tire chains are visible on these tires. A set of four tire chains were part of each M3A1's On Vehicle Material (OVM). (LOC)*

The observation commander in an M3A1 signals a halt during a mock battle at Fort Riley. Note the tire chains and the "B/arrow/5" markings in white on the left bumper. (LOC)

Above: *The crew of an M3A1 is poised for action during field exercises. The tall antenna is restrained by a thin piece of strapping. Note the small tactical sign of a downward-pointing arrow appearing on both the right rear bumper and to the rear of the door. (LOC)*

Below: *At Fort Riley, an M3A1 scout car operates within a smokescreen during a sham battle. The two visible crewmen are wearing fatigue hats with the brims turned up; the one in the front is training a Thompson submachine gun on a target. (LOC)*

Left: *A photo of an M3A1 commander consulting a chart reveals details of the machine gun mount. In this case, an air-cooled machine gun of the M1919 series, possibly an M1919A4, and its cradle-and-pintle assembly are mounted in an M30 carriage (sometimes called the trolley), the roller-equipped component that travels on the skate rail. The light-colored horizontal handle below the carriage is the track clamp handle. On the left side of the cradle is a .30-caliber ammunition tray. (LOC)*

Below: *A radio operator and his assistant, of Weapons Troop, 4th Brigade, 10th Cavalry Regiment, transmit messages from their M3A1 scout car at Camp Funston, Kansas, in December 1941. The photograph was taken from in front of the windshield, facing the rear of the vehicle. (NARA)*

Above: *Mexican officer students get a tour of an M3A1 scout car while visiting the Cavalry School at Ft. Riley, Kansas. They were part of a three-month course on modern warfare conducted there in the fall of 1941. This shot provides an excellent close-up of the right side of the vehicle. (NARA)*

Below: *A scout car leads two Ford GP jeeps through the desert at Ft. Bliss, Texas in the summer of 1941. The vehicles are taking part in 1st Cavalry Division maneuvers and the scout car, as part of the Blue Force, has taken the Red Force jeeps prisoner. (NARA)*

Left: *Here, an M3A1 speeds along a road at Camp Swift, Texas in 1942. The scout car shows the markings of the 95th Regiment, which at that time was completing its transition to become the Headquarters and Headquarters Battery of 95th Division's organic artillery unit. (NARA)*

Below: *Scout cars of the 107th Cavalry take the field alongside their four legged counterparts at Ft. Ord during maneuvers held there in May of 1942. The lead vehicle has a very unusual pattern on its tires. (NARA)*

Above: *Scout cars of the 107th conduct a driving exhibition while at Ft. Ord in May of 1942. Note the horseshoe-shaped insignia on each of the scout car's front bumpers. Large tents are visible in the background. (NARA)* **Below:** *Various types of vehicles, including several M3A1 scout cars in the foreground, were photographed on an LCT at the Amphibious Training Center, Camp Gordon Johnston, Florida on 18 February 1943. The M3A1s carry very little extra equipment. Only the center car appears to have a machine gun installed, a .30-caliber M1919A4, and even the machine gun tripods appear to have been left behind. (NARA)*

Above: *This M3A1 was photographed on 1 May 1943 at Camp Carson, Colorado. Its hoods are open. This vehicle was the scout car of the Reconnaissance Platoon Commander, 811th Tank Destroyer Battalion. Displayed in front of the vehicle is its On-Vehicle Material (OVM), comprising tool boxes, hydraulic jack, tire chains, pioneer tools, hand saw, liquid containers, weapons (including an M1903 rifle, an M1 bazooka, and a 30-caliber machine gun on tripod), and other equipment. (NARA)*

Left: *Another view shows the arrangement of the crew and stored equipment. Two men armed with an M1 bazooka and rifle sit facing each other to the rear of the radio rack. A gunner mans the M2HB .50-caliber machine gun, while the observation commander stands to the right. This is a later-production vehicle, with combat rims and racks on the cowls for 5-gallon liquid containers. (NARA)*

Above and below: *Although found in the Detroit Arsenal photo files with a 19 February 1943 date, evidence suggests that this vehicle was actually photographed at Aberdeen Proving Ground. Depicted is Lot 2 M3A1 scout car, U.S. Army registration number 601944. Stenciled on the side of the vehicle is the Ordnance Serial number 620. This vehicle, which was part of a 1,054-car order on purchase order T-278, was built in November 1940. Judging from the mismatched tread pattern of the tires, as well as the combat wheels (Lot 2 cars were built with Budd wheels), suggest that this vehicle was being used in a series of tests investigating tire wear, which was of great concern with regard to these vehicles early in the war. (NARA, both)*

Above: *A view inside an M3A1 set up as a command vehicle shows the orientation of the radio rack with the antenna mast mounted on top. Another mast is mounted to the front of it. An M2HB .50-caliber machine gun is at the center front of the skate rail. At the lower left is an M1919A4 .30-caliber machine gun, which has an ammunition tray with a beveled bottom attached to its cradle. Note the brackets holding the skate rail to the body. (NARA)*

Below: *A view from the front of the same command M3A1 shows the cramped interior. Two radio sets are crowded in between the ammunition lockers. Each is housed in a rack with a zippered, waterproof cover. On the left is a BC-312 HF-band radio receiver, and on the right is a transmitter. At the bottom are back-to-back passenger seats. At the rear, the .30-caliber machine gun is emplaced in a Model D36960 cradle-and-pintle assembly on an M30 carriage. (NARA)*

Above: *An M3A1 scout car loaded with cargo offered very little spare room for its crew. This example, at the Desert Training Center in 1941, is piled with cartons, ammunition boxes, and wooden crates. The armored windshield cover is lowered, with the sliding visor plates open. Protruding at the center is an antenna mast. (NARA)*

Below: *The same crew compartment, photographed from the front. Note the M3 37mm antitank gun hitched to the rear. At the lower left is a radio set. On the right are wooden crates for 37mm ammunition, to the rear of which are boxes of .50-caliber ammunition. Two back-to-back, side-facing passenger seats are at the rear. (NARA)*

Above: *This evaluation photo was taken at the Desert Training Center in 1941. Segmented poles have been placed along the sides to provide scale. This scout car has at least two different tread patterns visible, including a civilian style on the front axles. A spare tire is mounted on the rear, along with a rack for 5-gallon liquid containers. (NARA)* **Below:** *This M3A1 at the Desert Training Center is testing out a new recovery tool, a latticed mat. All of its weapons have been secured with canvas covers to keep out the dust and a woven screen has been attached to the radiator cover for the same purpose. (NARA)*

Above: *The U.S. Marine Corps acquired a number of M3A1 scout cars, including this one. It bears the insignia of the 1st Marine Division, the number 1 inside a diamond. On the hood is the USMC registration number, SC-7. The letters SC are an early registration-number code standing for "scout car." (NARA)*
Below: *On this USMC M3A1 scout car the armored windshield cover is lowered and the armored radiator shutters are closed. (NARA)*

Above: *A Marine Corps M3A1, registration number SC-8, drives off the ramp of a small landing craft during training exercises. Visible armaments include an M1917A1 .30-caliber machine gun and M2HB .50-caliber machine gun. (NARA)*

Below: *A USMC M3 tank and scout car during training at Tutuila, American Samoa during the summer of 1942. (NARA)*

Above: *USMC M3A1s parade down the streets of Melbourne Australia in May of 1943. This parade was part of the traditional "Down Under" celebrations held that year.* **Below:** *Ordnance workers in Algeria secure a damaged jeep to the top of an equally damaged scout car for evacuation to the rear. For this work an M980 Diamond T tractor has been hitched to a 40-ton Mk. 2 trailer—both no doubt borrowed from the British. The damaged material from the area will be moved via railroad to Algiers on this sunny North African day in February of 1943. (NARA, both)*

Above: *Lieutenant General George S. Patton was in command of II Corps when he was photographed in his personal, customized M3A1 scout car during the advance on Gabes, Tunisia, on 15 March 1943. The vehicle is fitted with an antenna mount on the right side of the body, a pedestal-mounted .50-caliber machine gun, and an armored shield for a forward-firing .50-caliber machine gun. (NARA)*

Right: *Patton's M3A1 also has an improvised liquid container rack that apparently could hold two 5-gallon cans. Fitted over the rear of the crew compartment is an armored awning. (NARA)*

Above: *An M3A1 scout car rushes past the remains of a Tiger I near Beja Tunisia on 26 April 1943. This was the location of the famous "Tiger graveyard" of demolished tanks of the German 501st Heavy Tank Battalion. U.S. units were rapidly advancing to Mateur, Tunisia and the end of the North African campaign.* **Below:** *Crated scouts cars are reassembled on the docks of Algiers on April 17, 1943. It was much easier to prepare the vehicles dockside, rather than move them to another location. In the background the unloading continues. All this equipment is being provided to reequip the French army prior to their departure to Italy. (NARA, both)*

Above: *These M3A1 scout cars are in French service. They are part of the newly formed Free French force being assembled in North Africa in 1943. U.S.A. numbers 6014502, 6014130, and 6014144 are visible. (Reg Hodgson collection)*

Below: *Another M3A1 scout car in French service in Italy bears the identification number 2157 F on the side of the body. On the rear of the body is an equipment rack fashioned from metal strapping. (Reg Hodgson collection)*

Above: *A White scout car of the 2nd Battalion, Grenadier Guards, 5th Guards Armoured Brigade, Guards Armoured Division, photographed on March 3, 1942. (IWM)* **Below:** *A scout car and a specially constructed 1-ton trailer connected and ready for travel. The full tarpaulin cover is in place. The very late style cover has clear plastic inserts in the rear sections. This photo was taken at Oakley Farms in Cheltenham, England on 22 December 1942. (NARA)*

An M3A1 scout car in Canadian service is being hoisted to or from a transport ship by means of netting slings under the tires. The car has an equipment rack on the rear and a registration number prefaced by the letters CZ. (Reg Hodgson collection)

Above: *The registration number CZ4258160 has been very roughly applied to the hood of this Canadian M3A1. Faintly visible below that number is the former U.S. registration number, 6090693. The front roller is absent.* **Below:** *The same M3A1, CZ4258160, is viewed from the rear. The registration number on the rear of the body is decidedly out of level. On the rear of the body is a spare-tire carrier, a common Canadian modification. (Reg Hodgson collection, both)*

Above: *Another Canadian M3A1 scout car bears registration number CZ4258169. An elaborate camouflage scheme of dark paint has been sprayed onto the canvas top as well as the lower part of the body. (Reg Hodgson collection)*

Below: *The same M3A1, CZ4258169, is seen from the other side, showing details of the camouflage scheme. Even the battery box on the running board has received the two-color treatment. (Reg Hodgson collection)*

Above: *A soldier lowers the swing arm of a spare-tire carrier on a Canadian M3A1. The rear of the bodies of Canadian M3A1s had this carrier and lacked the rear bumper and the tripod brackets. (Reg Hodgson collection)*

Below: *This British M3A1 has been outfitted as an ambulance. Two patients are on stretchers suspended in the crew compartment and three wounded men are sitting in the center of the crew compartment. (Patton Museum)*

Above: *An immaculate White scout car carrying His Majesty, King George VI, inspecting the British 6th Armoured Division at Lakenheath in Suffolk in 1941. (The Tank Museum)* **Below:** *Elements of the Fifth Army in San Giorgio, Italy, make preparations to attack their next objective, Esperia, on 16 May 1944. At the front is an M3A1 scout car. Note the flattened headlight and brush guard on the right fender, and the large, white diamond tactical symbol on the side of the body. (NARA)*

Above: *British troops and members of the French resistance shelter behind a White scout car as an ammunition truck burns in the distance after being hit by shellfire on 26 July 1944. Note the rear step designed to hold the tripod mounting brackets has been removed and different military tail lights added. This was a common modification in the British Commonwealth armies. (IWM)* **Below:** *Major-General "Pip" Roberts of the British Army is seen commanding 11th Armoured Division in NW Europe from his White scout car on 15 August 1944. The skate rail has been removed from this car, no doubt to improve egress. (IWM)*

Above: *A White scout car and other command vehicles enter the town of Lannoy in France, 6 September 1944. (IWM)*
Below: *Parked outside a private address, probably somewhere in Belgium, a well-laden White waits at the side of the road. This car was assigned to the 11th Hussars, which was attached to the 7th Armoured Division. (The Tank Museum)*

Above: *An M3A1 scout car of the Princess Irene Brigade enters The Hague in the Netherlands on 9 May 1945. The brigade was a military unit initially formed from approximately 1,500 Dutch troops, including a small group guarding German prisoners-of-war, who arrived in the United Kingdom in May 1940 following the collapse of the Netherlands. (Image Bank WW2)* **Below:** *Rows of surplus M3A1 scout cars await sale by the U.S. Treasury Department at the Lordstown Ordnance Depot, Warren, Ohio. These were among the 2,215 scout cars the Army had recently declared surplus. Although the M3A1 scout car was becoming obsolete when the United States entered World War II, the vehicle saw a fair amount of combat in most theaters. (NARA)*

Chapter 3 Personnel Carrier T14

A Design Takes Shape

Following several successful experiments with an M2A1 scout car equipped with half-tracks, the Mechanized Cavalry Board recommended further development of this type of vehicle. The vehicle desired by the Board was to be substantially an M3A1 scout car with the rear wheels replaced by tracks. Cavalry Board recommendations found concurrence from the Chief of Field Artillery. Following up on these recommendations, on 26 December 1939 the Artillery Division, Industrial Service, Ordnance Department delivered drawing D-42876 laying out such a vehicle. The Ordnance Committee, through action OCM 15544, authorized the manufacture of a pilot vehicle, designated T14, on 28 December 1939. White Motor Company of Cleveland, builder of the M2A1 and M3A1, was the natural choice to build the new design, and was awarded the contract to produce the vehicle.

To speed the process production of the pilot model and subsequent testing process, the 10-inch wide track, rear-drive suspension along with the front wheel driveline would be the same as that used on the Half-track Personnel Carrier T7. These components were used despite the new specification calling for 12-inch wide tracks and front-mounted drive sprockets. The vehicle was powered initially by a White model 20A 6-cylinder, 116 horsepower gasoline engine developing 280 foot-pounds of torque.

The new vehicle was designed to carry a payload of 3,400 pounds at 40 MPH. With body armor of ¼-inch plate and ½-inch windshield armor, the vehicle had a gross weight of 15,000 pounds. Armament was two .30-caliber machine guns and one .50-caliber machine guns, all mounted on a common skate rail encircling the crew compartment. Tripods were strapped to the rear of the vehicle, allowing the machine guns to be relocated to ground mounts. Crew entry and exit was either over the sides or through the doors for the driver and codriver. Like the M3A1 scout car, no rear door was provided.

Once the T14 was completed, it was driven from White's Cleveland plant to Aberdeen Proving Ground, Maryland. Departing Cleveland on 28 May 1940, the vehicle arrived at the Proving Ground the next day, where it underwent testing through 28 September 1940.

During the course of the Aberdeen tests, four further powerplants were evaluated in the vehicle. The additional engines tested were the Hercules WXLC3, White 140A, White 160A and the Buick Series 60. The 142-horsepower Buick and 147-horsepower, 386-cubic inch White 160A provided the best performance, with the 160A's abundant torque being a decided advantage.

Above: *The T14 half-track scout car was the pilot for the M2 and M3 half-tracks. Built in early 1940 by the White Motor Company, Cleveland, Ohio, the vehicle was tested at Aberdeen Proving Ground, Maryland, from May to September 1940. Several six-cylinder engines were tested in the T14, with the Buick Series 60 and White 160A yielding the best horsepower and torque. (NARA)*

Below: *Notable features of the T14 include doors for large storage boxes to the rear of the side doors and a body that ends above the center of the idler wheel. The rear of the body is ten inches shorter than that of the M3 half-track personnel carrier. In addition, the drive axle is at the rear of the vehicle. (NARA)*

U.S.A.W-601128

U.S.A.W-601128

Above: *The design of the idlers and drive sprockets differ from those of the standardized vehicles, and the T14's bogie wheels (called "rollers" in technical manuals) were open-spoke. The T14's bogie bracket, suspension arms and frames, and roller support are also different in design from the production versions. In addition, the T14 lacked the prominent idler springs of the standardized half-tracks.* **Below:** *A roller assembly with shock-absorbing springs was tested on the front bumper of the T14 and was found to improve the vehicle's mobility over ditches. Note the crossed-sabers cavalry placard below the lowered upper panel on the side door. (NARA, both)*

Above: *The bogie assembly of the T14 includes a bracket assembly with an upper track roller on top (center); two vertical volute springs within the bracket (out of view); two suspension arms pivoted at the bottom of the bracket; and suspension frames, to which the bogie wheels are attached, pivoted to the ends of the suspension arms. Levers termed "crabs" are linked to the volute springs, with the outer ends of the crabs resting on top of the suspension arms. (NARA)*

Right: *As originally configured, the T14's bogie wheels lacked the rubber tires of the standardized M2 and M3 families of half-tracks. During the latter stages of testing in December 1940, rubber-tired wheels were mounted on the vehicle. The shaft to the right was termed the bogie cross tube; it serves to strengthen the bogie assemblies. Standard half-tracks have a yoke-type mount for the upper rollers instead of the single, inboard bracket shown here. (NARA)*

Above: *A view of the inner side of the T14 bogie assembly. Above the bogie cross tube is the flange of the bracket, which is bolted to the chassis.* **Below:** *This view from below of the center of a bogie assembly illustrates how the suspension arms are pivoted to the bottom of the bogie bracket. The crabs are hanging loose below the arms. (NARA, both)*

Above: *An overhead view of the T14 bogie assembly, showing the cross tube connecting the two bogie units. (NARA)*

Right: *Powering the T14 was the White 160A engine, a six-cylinder, 386 cubic inch, four-cycle, in-line model. It was capable of 147 horsepower at 3,000 rpm and had a maximum net torque of 325 foot-pounds at 1,200 rpm. To the rear of the engine is a White transmission. (NARA)*

Above: *Although the forward part of the T14 was similar to the M3A1 scout car, the similarities ended to the rear of the cabs, with the top of the body of the T14 having a straight profile from the top of the front door to the rear of the body, as opposed to the M3A1's sloping profile. (NARA via Jim Gilmore)*

Below: *Like the M3A1 scout car, the T14 half-track had provisions on the rear of the hull to stow three machine gun tripods: two for the .30-caliber guns and one for the .50-caliber. Mounted on the .30-caliber machine-gun tripods are cradle assemblies. (NARA via Jim Gilmore)*

Above: *The T14 was designed as an artillery prime mover and had two ammunition lockers to the rear of the cab seats; each locker had an exterior door. A rear-facing seat was between the two lockers, and there were three inward-facing crew seats on each side of the body. (NARA via Jim Gilmore)*

Below: *This elevated view of the T14 with a M2A1 105mm howitzer hitched to it was taken at Aberdeen Proving Ground on 10 December 1940. In contrast to the front wheels shown in the first several photos in this chapter, the front wheels here are a solid-disk type. (NARA via Jim Gilmore)*

Chapter 4
G-102 Half-track Production

Autocar, White and Diamond T

Although the initial testing of the T14 was not 100% successful, largely owing to the reuse of the T7 components and the early, underpowered engine, enough promise had been shown that the Chief of Field Artillery wrote to the Chief of Ordnance on 8 August 1940 requesting that half-track vehicles be procured as prime movers and artillery carriers.

Acting on this request, the Ordnance Committee on 19 September 1940, OCM 16112 recommended that the T14, with the White 160A engine and other modifications as dictated by testing, be classified Standard as the M2 Half-track Car. Plans were initiated and specification AXS 486 written for the M2, with intended dual roles of scouting and artillery prime mover. Two large stowage compartments, one on each side, were located just behind the vehicle doors. The tops of these stowage compartments could be opened from inside the vehicle, providing access to the top shelf within, while opening doors on the outside of the vehicle provided access to the shelves throughout the compartment.

Inside the vehicle, beyond the driver's and co-driver's seats, were further accommodations for eight men. Near the interior rear of each side armor panel was a self-sealing fuel tank.

At almost the same time that the M2 was being developed for Artillery, Ordnance was pursuing their own idea of a half-track through work with the Diamond T Motor Car Company, with the intent to create armored personnel carriers. Known as the T8, this initiative was authorized by OCM item 15947. This vehicle would be standardized as the M3. The M2 and M3 vehicles were assigned Standard Nomenclature List (SNL) number G-102. The M2 body was designed for a capacity of 10 men, while the M3 had a capacity of 13 men. The M2 was intended as a prime mover for artillery up to a 155mm howitzer, and the M3, while originally intended as an armored personnel carrier, would come to be the basis for a wide variety of vehicles.

On 13 September 1940, production order C-515 was issued to Rock Island Arsenal, authorizing 424 Half Track Cars, M2. Invitations for bid to produce this vehicle were sent to a number of manufacturing firms, including White, Autocar and the Diamond T Motor Car Company. The Autocar Company of Ardmore, Pennsylvania was the lower bidder, and was awarded the first production contract for the vehicles. This contract, 741-ORD-6276, for 424 of the new vehicles, was awarded during September 1940. This was later increased to 537 vehicles.

The new vehicles were in high demand, and additional sources of supply were needed. The presidents and executive vice-presidents of the three lowest bidders, Autocar, White and Diamond T, met with Colonel W. W. Warner, Chief of the Artillery Division, Industrial Service, Chief of Ordnance on 28 September 1940. As a result of this meeting, and a subsequent meeting at the Autocar plant on 1 October, it was agreed that all parts except armor plate produced by all three companies

The T14 undergoes trials at Aberdeen Proving Ground. Note the placard on the right headlight brush guard that reads, "APG-400." Surrounding the inside of the body is a skate rail for machine guns. Radio equipment, including two antennas, is also within the vehicle. (NARA)

APG-400

would be interchangeable, and the Half-track Engineering Committee was formed.

On 20 September 1940 Production Order C-540 was issued to Rock Island Arsenal authorizing 4,908 more M2 vehicles. In response, contract 741-ORD-6285 was issued to White Motor Company for the half-tracks—or, as White referred to them—Half-tracs. At this time there was no Army Supply Program, thus the Artillery Division was the originator of these procurements. These vehicles were associated with Production Order C-544. Additionally, on that same day orders were placed by negotiated contract with Autocar for 1,382 M3 half-track personnel carriers on Production Order C-544.

On 23 September 1940, Production Order C-545 was issued to the Arsenal. This authorized procuring 2,000 of the M3 models through negotiated contract with the Diamond T Motor Car Company. These were covered by contract W-741-ORD-6276.

By mid-October, with the details worked out and two production contracts having been issued, the Ordnance Committee approved the earlier recommendation and the M2 was classified as standard. The first vehicles of the type delivered were 62 units delivered by White in May 1941. In June the first five M2 vehicles were delivered by Autocar, as were four M3 vehicles. The first of the Diamond T-produced M3s was delivered in May 1941. The 537 Autocar-produced M2s were destined for foreign military sales, a role eventually taken over by the M9, discussed later in this book.

Ultimately, demand for the vehicle was such that White began production as well, with that firm's initial M3 contract, W-303-ORD-945, being awarded in July 1941. However, because White was the sole source for the M3A1 scout car, and already had a commitment to build M2 half-track cars, contract W-303-ORD-945 was for a modest 100 examples only.

The White model 160AX engine powered the M2 (and M3), driving through a Spicer model 3641 special combination transmission and transfer case. On some vehicles this unit was equipped with a power take off through which a front mounted winch was driven. A Timken model 56410-BX-67 rear axle drove the tracks, while the front was supported by a Timken F-35-HX-1 axle on 8.25-20 combat tires. The body was constructed of ¼-inch face hardened armor plate.

B.F. Goodrich developed the 12-inch wide band tracks with 58-inch pitch.

During 1941, 1,182 half-tracks were accepted from Autocar, 1,087 from Diamond T, and 3,242 from White. The following year 4,735 were accepted. In 1943 a further 3,115 M2 vehicles were accepted, making all time production of the M2 a total of 11,415, with 2,992 coming from Autocar and 8,423 from White.

As impressive as those numbers were, they were miniscule compared to the Army Supply Program (ASP) forecasts of the time. The 11 February 1942 Army Supply Program called for 21,938 half-tracks by the end of 1942, a total of 123,712 by the end of the following year, and a whopping 188,404 by the end of 1944. These quantities would be reduced in time, with the June 1942 ASP reducing the quantities to 21,653, 63,471, and 101,458 respectively. At that same time, White was directed to cease half-track production in order to increase M3A1 scout car production. That directive was rescinded shortly after issue.

The most visible change to the M2 and M3 vehicles came about as a result of a 19 May 1942 Ordnance Committee recommendation, suggesting a revised .50-caliber machine gun mounting was adopted. For the M2, this amounted to eliminating the skate rail and its M35, and sometimes M29 skate mounts, and using instead a M49 ring mounted above the assistant driver's seat. This concept was approved 25 June 1942 by OCM 18394. Two pilot vehicles, designated M2E6 were authorized. After testing of two prototypes of different styles, one of which included a moderate amount of armor protection surrounding the ring, the new mounting with armor was adopted as standard. On the M3, the change meant the elimination of the M25 Pedestal Mount and incorporating the M49 ring mount.

Above: *Engine blocks for half-track scout cars are stacked at the White Motor Company assembly line in Cleveland, Ohio. (LOC)*

Below: *A workman at White lowers a White 160AX engine block onto an assembly line. The studs and bearing caps have been installed on the block. (LOC)*

Almost a year later, on 6 May 1943 OCM item 20368 approved the recommendation of OCM 20070 of 1 April 1943, adopting the armored ring mount.

Moving ahead, this type of mounting would be used, and the vehicles so equipped were designated M2A1. A similar change was made to the M3, the new designation being M3A1. Enthusiasts have dubbed this arrangement a "pulpit" mount. No longer having the skate rail to which the .30 caliber weapons could be attached, these vehicles had three fixed sockets bolted to the rear armor, one on the rear and one on each side. A single M1919A4 was the standard .30 caliber armament for the M2A1, which entered production 1 October 1943. M2A1 acceptances in 1943 amounted to 987 vehicles, and in 1944 a further 656 were accepted.

In order to incorporate the new gun mounting into vehicles already in the field, a contract was awarded to the Miller Printing Machine Company of Pittsburgh to produce 6,000 ring mount modification kits. These were to be used to field modify M2 and M3 vehicles to A1 standards.

While this change to the M2 and M3 was readily apparent, it was not the only change to impact the G-102 series vehicles. Some of these changes were made after the vehicles were manufactured. These changes were implemented through a series of Modification Work Orders, or MWOs. Several of these are described in detail in later chapters.

Other changes were planned and, typically, made in the design and implemented on the production line. These were documented in a series of Issue Letters, and some of these are worthy of description. One example such change was the incorporation in production of the "Protectoseal" gasoline tank cap. This change, which addressed complaints from the field about water entering the tank through the previously used brass filler cap, was released through Issue Letters on 28 April 1943, 31 July 1943 and 23 October 1943. The change was also released to the field through MWO G102-37 on 15 July 1943.

Significantly, Basic Issue Letters were issued on 5 May 1943, 21 May 1943, 19 May 1943 and 10 May 1943 that summarized the characteristics of the M2A1, M3A1, M5A1 and M9A1 respectively. Further changes to these designs were covered in a series of Issue Letters. Engineering Change Orders followed up the Issue Letters. While most of these Issue Letters were implemented, it is worthwhile to mention that not all were—and that the delay between the date of Issue Letter and implementation could vary widely.

On 7 September 1943 an Issue Letter specified a reduction in the amount of sealant material used in the bulletproof gasoline tanks. This was part of an overall effort to reduce the use of crude rubber throughout all military equipment.

On 15 October 1943 a revision for the bogie slide was directed in order to increase strength and improve wear characteristics. Three days later further changes were directed for the suspension, this time focusing on improved bogie arms and rub plates.

The horn was the focus of an Issue Letter on 26 October 1943, directing a standard Ordnance horn be used. This was done to serve the two-fold benefit of increasing Ordnance standardization as well as providing a louder horn. The next day an Issue Letter was issued directing the incorporation of Ordnance standard taillights. This provided resilient mounting as well as standardization of this component.

The various manufacturers used their own internal systems to document and disseminate the changes called for by the issue letters. White used a series of service bulletins. For example, White Service Bulletin 115 of 31 March 1943 impacted the M2, M3, M4 and T30 vehicles, listing new part numbers for the bodies of the vehicles. It recorded "Nature of change—New bodies to include Pioneer Tool brackets and 30 caliber and 50 caliber tripod brackets" as well a "Reason for change—White to furnish Pioneer Tools, Pioneer Tool Brackets and tripod brackets by Diebold."

The adoption of the Protectoseal filler caps, described in the Issue letters named above was implemented through White Service Bulletin 120, on 1 May 1943. This bulletin specified that the change

A weights and standards technician at White carefully weighs connecting rods to make sure they fall within the tight tolerances necessary for smooth engine operation. (LOC)

be made at serial number 278144 on 20 March 1943. The bulletin further provides that the new White part number for the filler cap is 413110, replacing the old part number 387278, and lists the authorization per change order for each contract. Of note, the date this change was implemented by White does not correspond with the dates of any of the Issue Letters, indicative of the lag between decision and implementation of these type actions.

Service Bulletin 112 of 31 March 1943 reveals that at serial number 250280 the instrument bezels where changed from chrome to olive drab, and at the same time the gauges were cemented dust-tight.

In mid-1943 the Military Standard oil filter began to be used on half-track engines. This was part of an effort to standardize oil filters on combat vehicles. White service bulletin 182, of 4 March 1944 documented this change, recording that the change went into place per change order AE on contract W303-ORD-1611 and change order Z on contract W303-ORD-2080.

In early January 1944 the Office Chief of Ordnance in Washington requested that a production plan be submitted concerning the manufacture of 10,089 half-tracks from 1 January 1944 through 31 December 1945. On 27 January 1944 the Director of Material General Staff Corps established production goals of 2,673 in 1944 and 7,032 in 1945.

The manufacturers, however, had already scheduled production in January and February of 1944 greater than the new requirement for the year. On 29 January a meeting was held in the offices of the Chief of Ordnance Detroit, with representatives of White, Autocar, Diamond T, and the Cleveland, Philadelphia and Chicago Ordnance Districts. A plan came out of that meeting to complete the entire 1944 authorization by 31 March 1944, and cancel any remaining vehicles outstanding from previous forecasts, a total of 22,655 vehicles. After this cancellation, White asked for and was given permission to produce a further 182 vehicles, as production had advance too far to permit economical cancellation. In total, 7,371 White vehicles were cancelled, along with 7,527 Autocar and 7,575 Diamond T. The existing spare parts orders were allowed to stand, and were augmented with an order for a further 300 sets of spares, these amounts representing an all-time buy of spares intended to support these vehicles through the remainder of their service life as well as supporting a rebuild program.

Despite these efforts, by 1951 the U.S. Army had contracted with White for additional replacement engine production. These engines were shipped in the latter part of the year.

Production of half-tracks of all types during 1944 amounted to 1,115 by White, 915 by Autocar and 825 by Diamond T.

However, the Army Supply Program of 1 February 1944 also listed a requirement for 5,250 Standard Half-tracks (and 2,895 International Harvester models). The facilities that had previously manufactured half-tracks were all asked to participate in the remanufacture, although the quantity initially forecast had changed. Ultimately, the remanufacture of 4,712 half-tracks was offered in a three way split to White, Autocar and Diamond T. White immediately refused to participate, citing other government commitments, and the 1/3 originally allocated to White was divided between Diamond T and Autocar. At about this time, the decision was made to have 853 of the vehicles remanufactured in Service Command shops. Additionally, Bowen and McLaughlin of Phoenix, Arizona, was awarded the remanufacture of 400 half-tracks.

Diamond T, however, elected to not participate in the program, leaving the Chicago Ordnance District to contract with E.W. Wylie of St. Paul, Minnesota to remanufacture 655 of the vehicles originally allocated to Diamond T.

Further changes to these allocations were made as well. *The Summary Report of Tank-Automotive Material Acceptances* listed the following acceptances of rebuilt half-tracks as of 1 September 1945:

Above: *A mechanic inserts pistons into the cylinders of a White 160AX engine. According to this photograph's original caption, this man was a former auto worker. (LOC)*

Below: *Workers adjust a flywheel housing. Note the rear engine supports on either side of the housing. (LOC)*

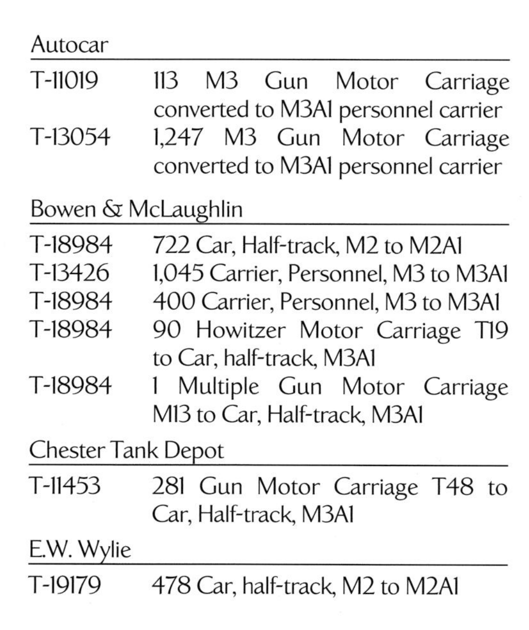

Autocar

T-11019	113 M3 Gun Motor Carriage converted to M3A1 personnel carrier
T-13054	1,247 M3 Gun Motor Carriage converted to M3A1 personnel carrier
T-13146	595 Carrier, Personnel, M3 to M3A1
T-19179	169 Carrier, Personnel, M3 to M3A1
T-19179	1 Howitzer Motor Carriage T30 to Car, Half-track, M3A1

Bowen & McLaughlin

T-18984	722 Car, Half-track, M2 to M2A1
T-13426	1,045 Carrier, Personnel, M3 to M3A1
T-18984	400 Carrier, Personnel, M3 to M3A1
T-18984	90 Howitzer Motor Carriage T19 to Car, half-track, M3A1
T-18984	1 Multiple Gun Motor Carriage M13 to Car, Half-track, M3A1

Chester Tank Depot

T-11453	281 Gun Motor Carriage T48 to Car, Half-track, M3A1

E.W. Wylie

T-19179	478 Car, half-track, M2 to M2A1

Richmond Tank Depot

T-11027	5 Car, Half-track, M2
T-11347	61 Car, Half-track, M2 to M2A1

White Motor Company

T-3844	108 Howitzer Motor Carriage T30 to Car, Half-track, M3A1

The Ordnance Department report Engineering of Half-Track vehicles, 1 June 1945 to 15 September 1945, recorded "The demand for half-track vehicles, previously met by remanufacturing the required vehicles after production was discontinued, dropped sharply and remanufacturing was terminated in June 1945."

Above: *The workman at left adjusts the carburetor of a White 160AX engine at the White factory. The vertical cylinder toward the front of the engine is the oil filter. (LOC)*

Below: *Engines make their way down the White assembly line. Note the parts trays and boxes behind the worker at the right. (LOC)*

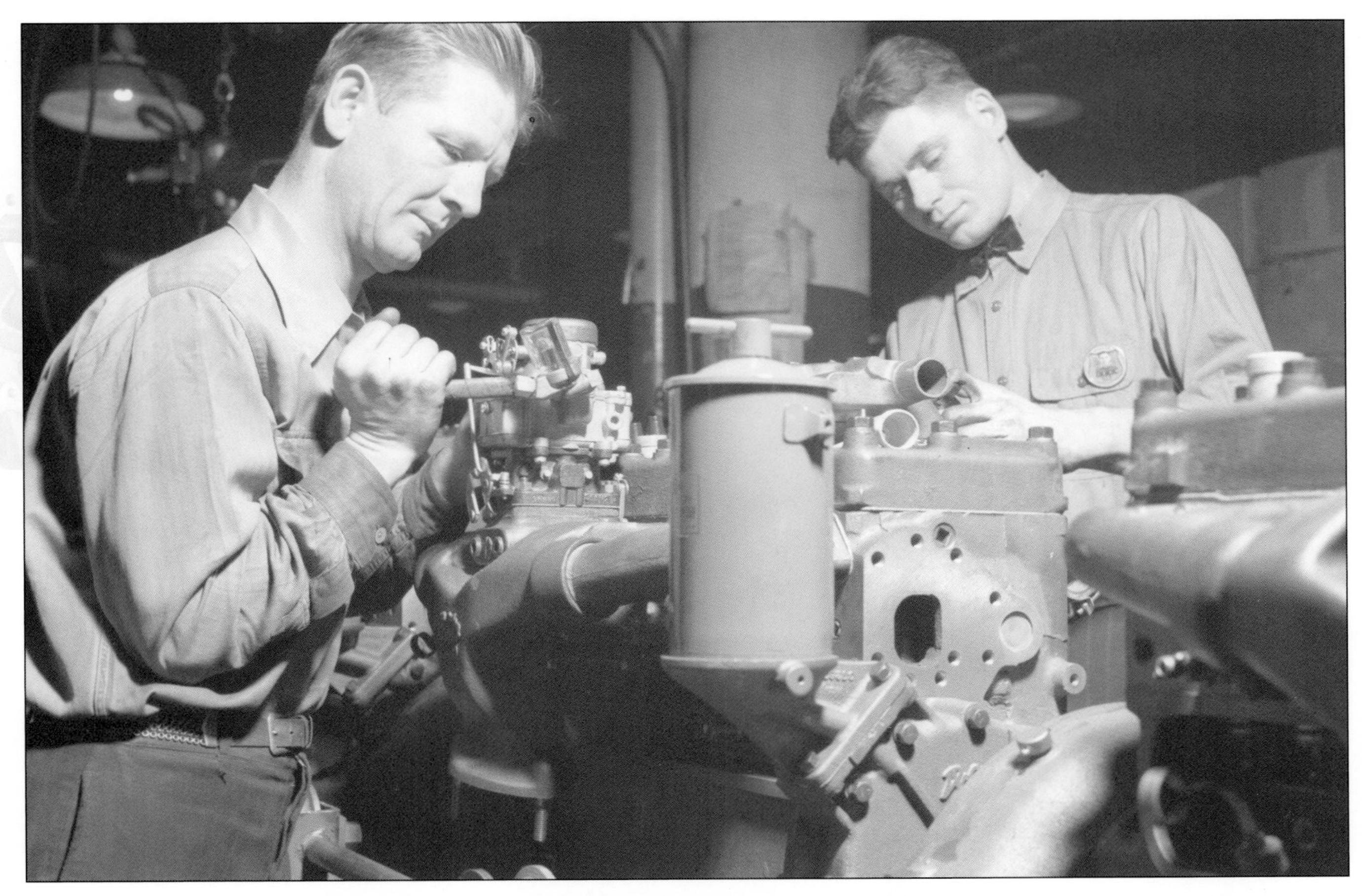

Assembly-line workers at White install the ignition wires on 160AX engines. The front engine mount or trunnion and the circular vibration damper are visible toward the bottom front of the closest engine. (LOC)

Above: *Technicians at the White Motor Company test a 160AX engine with a dynamometer to check its power. The fan is not installed on the engine. A flex-hose carries exhaust fumes safely away. (LOC)*

Below: *White Motor Company employees roll two finished 160AX engines down a track. Note the trolleys on which the engines rest. The "White" script logo and a casting number are on the front of the engine block. (LOC)*

A worker at the White plant uses an electric hoist to lift the engine for a half-track. At this late stage of the production process, the fan has been installed. (LOC)

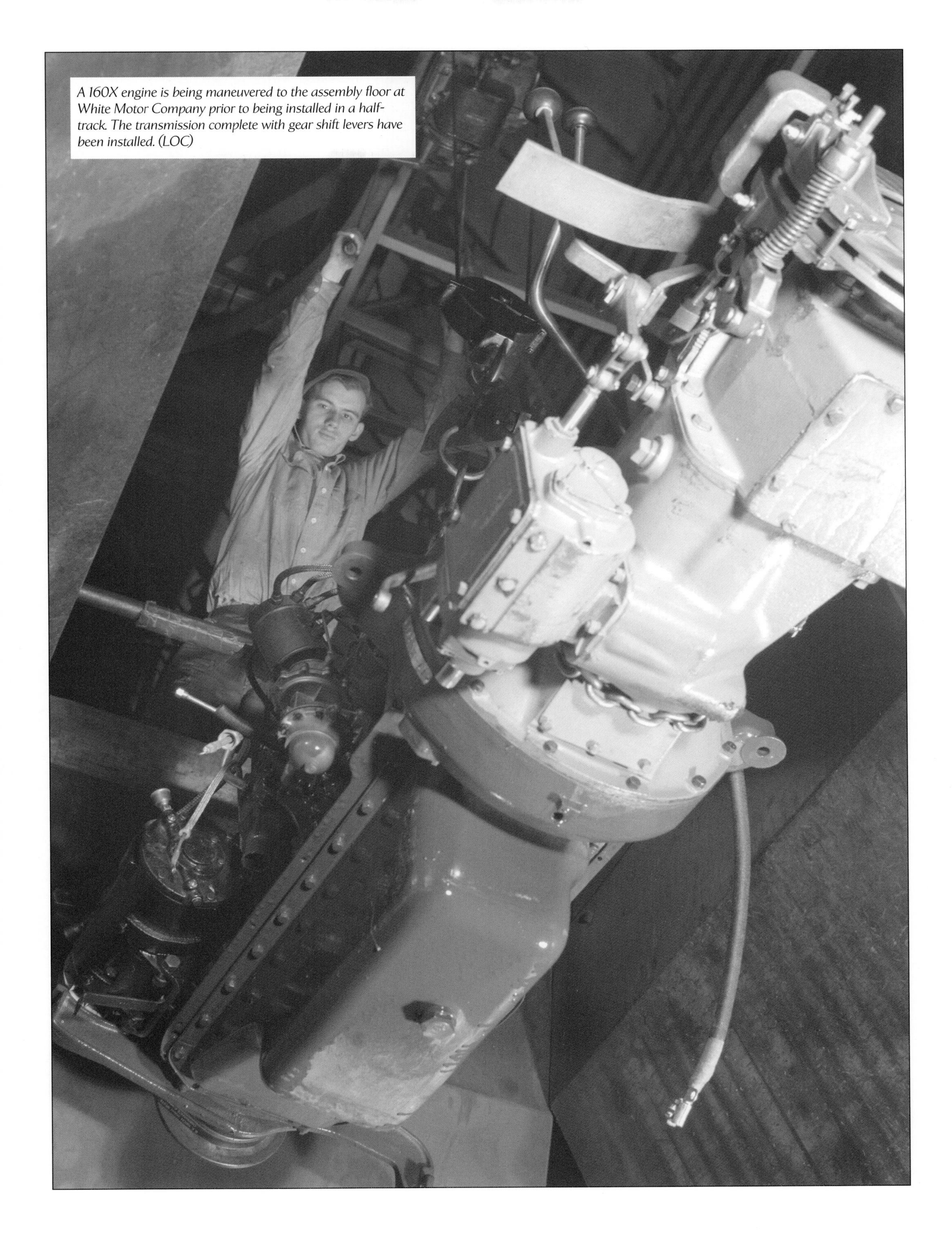

A 160X engine is being maneuvered to the assembly floor at White Motor Company prior to being installed in a half-track. The transmission complete with gear shift levers have been installed. (LOC)

Above: *Mechanics at White wrestle an engine onto the chassis of a half-tack. The winch-equipped front bumper of the half-track is in the right foreground. (LOC)*

Left: *A factory photo of a nearly complete White 160AX engine, minus the transmission. The generator is on the lower forward part of the side of the engine block, while the distributor, cranking motor, and fuel and vacuum pump are to the rear of the generator. The air cleaner (an atypical type is shown here) sits atop the carburetor. (LOC)*

Right: *The left side of the engine from the rear. The painted White logo is partially visible below the ignition coil. The coil is installed in a late style bracket. The mechanism below the distributor is the oil temperature regulator. Over the generator is the water pump. (LOC)*

Below: *A forklift operator loads or unloads partially completed engine assemblies. Note the square protective covers, possibly made of cardboard, fitted over the three studs on the exhaust manifolds. (LOC)*

Above: *Two workmen at White Motor Company are preparing bogie assemblies for painting. Stacked up behind them are scores of front-wheel rims and idler wheels.* **Below:** *The mechanic at left mounts suspension frames on bogie wheels using an impact wrench, while the worker to the right prepares to lower a bogie bracket assembly in order to attach the bogie wheel/suspension frame assemblies. Note the two vertical volute springs in the left bogie bracket. (LOC, both)*

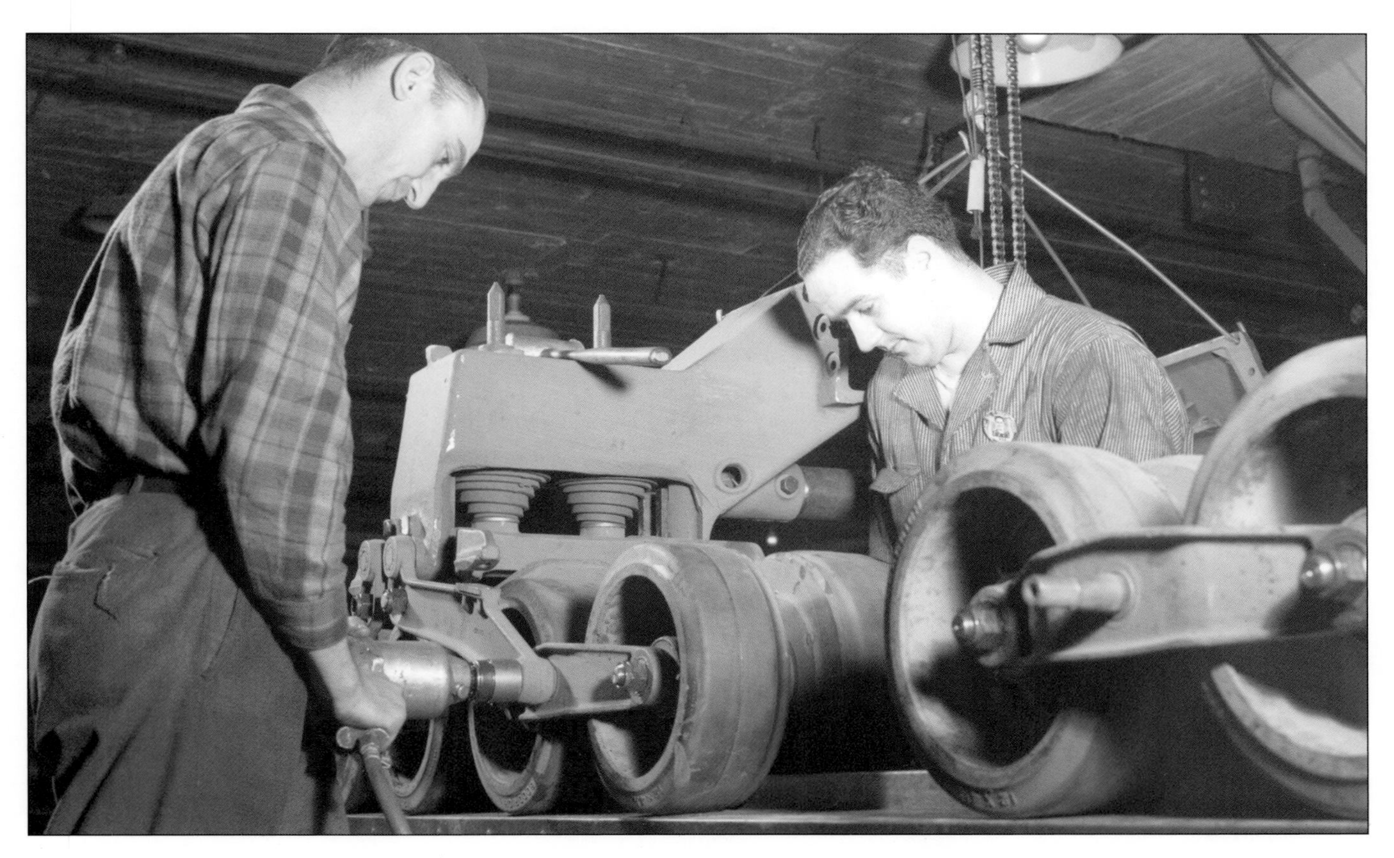

Above: *The bogie bracket assembly has been lowered into place, and the mechanic tightens a nut securing the suspension arm to the suspension frame. Note the mold seam around the center of the rubber tires.* **Below:** *An inspector at the White Motor Company plant touches up a part on a bench grinder amid piles of idler wheels. (LOC, both)*

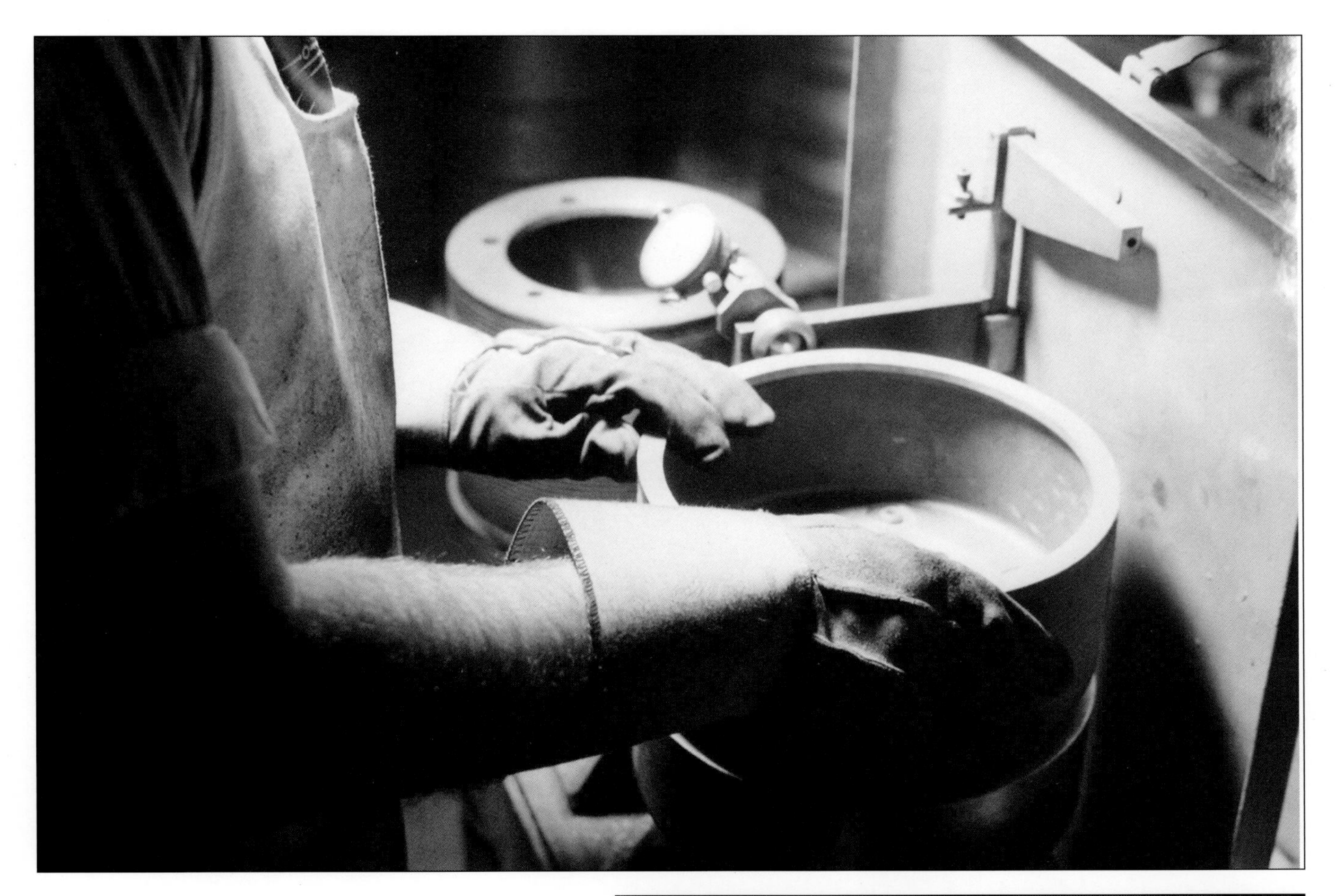

Previous page, top: *A painter sprays a partially completed half-track chassis. Lying atop the rear of the chassis are the upper track rollers, which are not yet mounted on the bogie brackets. The towing pintle is installed on the rear of the chassis. (LOC)*

Previous page, bottom and above: *An inspector checks the steel rims of bogie wheels at the Goodrich factory, Akron, Ohio. (LOC)*

Right: *A mechanic at White Motor company installs rubber tires on bogie wheels. (LOC)*

Above: *The tracks for the M2 and M3 families of half-tracks have an endless-band structure, rather than individual track shoes with removable end connectors. Steel cables run around the interior of each track assembly, providing lengthwise strength. Steel crossbars are bolted to the cables at intervals. At the centers of the crossbars are guide plates that the drive sprockets engage to propel the tracks. The cable/crossmember assemblies are encased in vulcanized rubber, forming continuous tracks.* **Below:** *Workers assemble the metal parts for a track on a form called the "building wheel" at the Goodrich plant in Akron, Ohio. (LOC, both)*

Above: *This morale-boosting poster hung in the Goodrich plant during World War II.* **Below:** *Two tracks for a half-track have just come out of the hydraulic curing press at Goodrich and are almost complete. (LOC)*

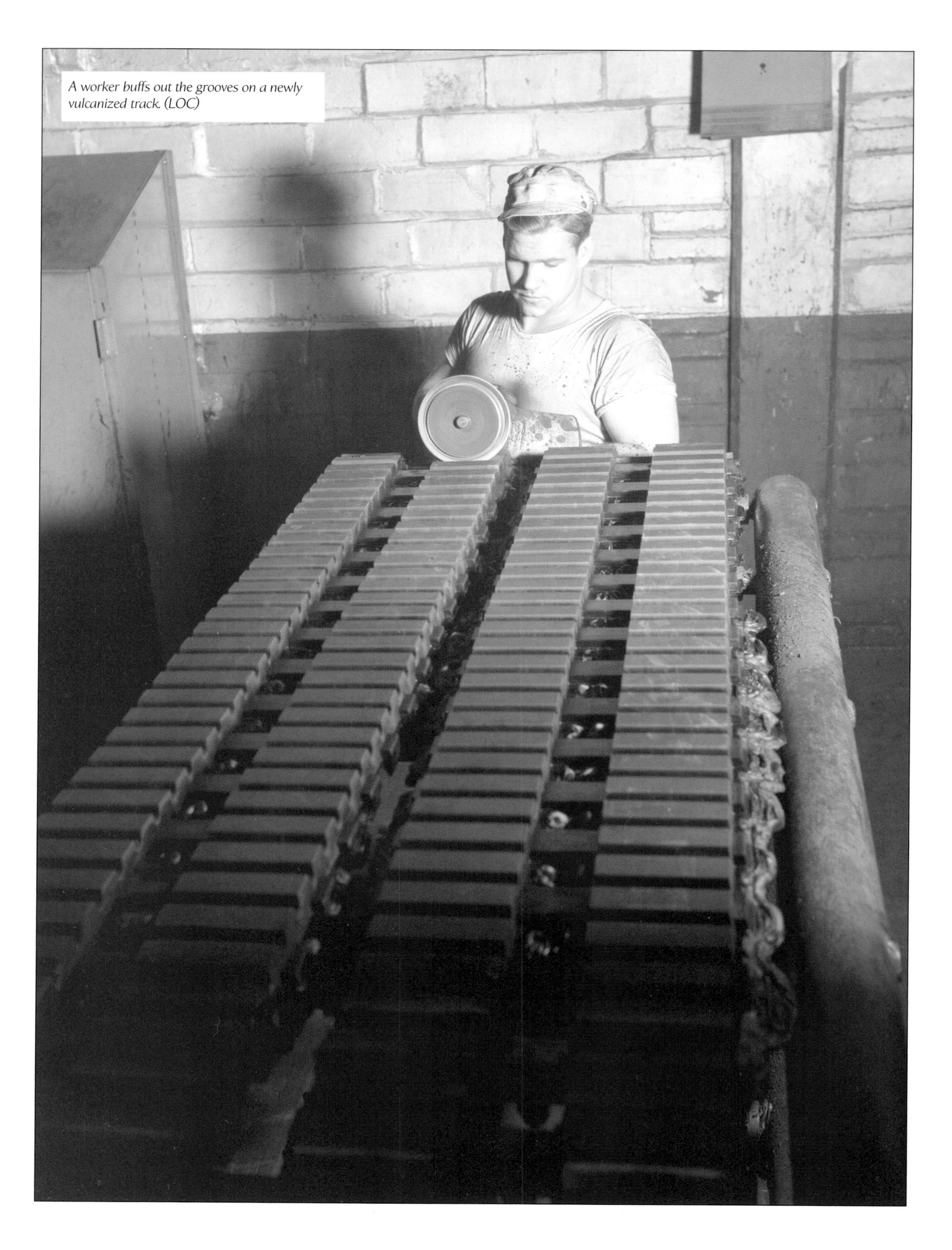

A worker buffs out the grooves on a newly vulcanized track. (LOC)

Another worker at Goodrich puts the finishing touches on a track assembly on the trimming rack. He is removing the curing plates and rubber overflow. (LOC)

A worker mounts the center guides on the crossbars on a track assembly at the Goodrich plant. The structure on which the track assembly is mounted was called the finishing machine. (LOC)

At the Goodrich factory, a worker manipulates a chain hoist in preparation for moving a track assembly fresh out of the curing press to the trimming racks, where the curing plates and rubber overflows will be removed. (LOC)

Two inspectors at the Goodrich factory carefully examine tracks prior to shipping. (LOC)

Above: *An inspector makes a final check of serial numbers on tracks packed for shipping from the Goodrich plant. (LOC)*

Below: *Four workers at the White factory wrestle a track onto the right side of a half-track chassis. Behind the man to the right is a large, multi-tiered parts bin. (LOC)*

Above: *Another view of three of the same men manipulating the track into place on the left side of a half-track. The outer wheel of the idler has been removed to enable track installation, and will be attached to the inner wheel once the track is in place.* **Below:** *Using a wooden slat as a pry bar, a White employee mounts the left wheel on a half-track. Note the headlight brush guard on top of the fender. (LOC, both)*

Above: *A worker at left checks the upper track roller, while another mechanic makes adjustments in the driver's compartment. The instrument panel is already in place, but the map compartment door is not installed yet. (LOC)*

Below: *Half-track chassis are on the White assembly line prior to the installation of bodies. The drum-shaped mechanism to the front of the worker at the left is the Bendix 370001 vacuum booster, used on early production G-102 vehicles. (LOC)*

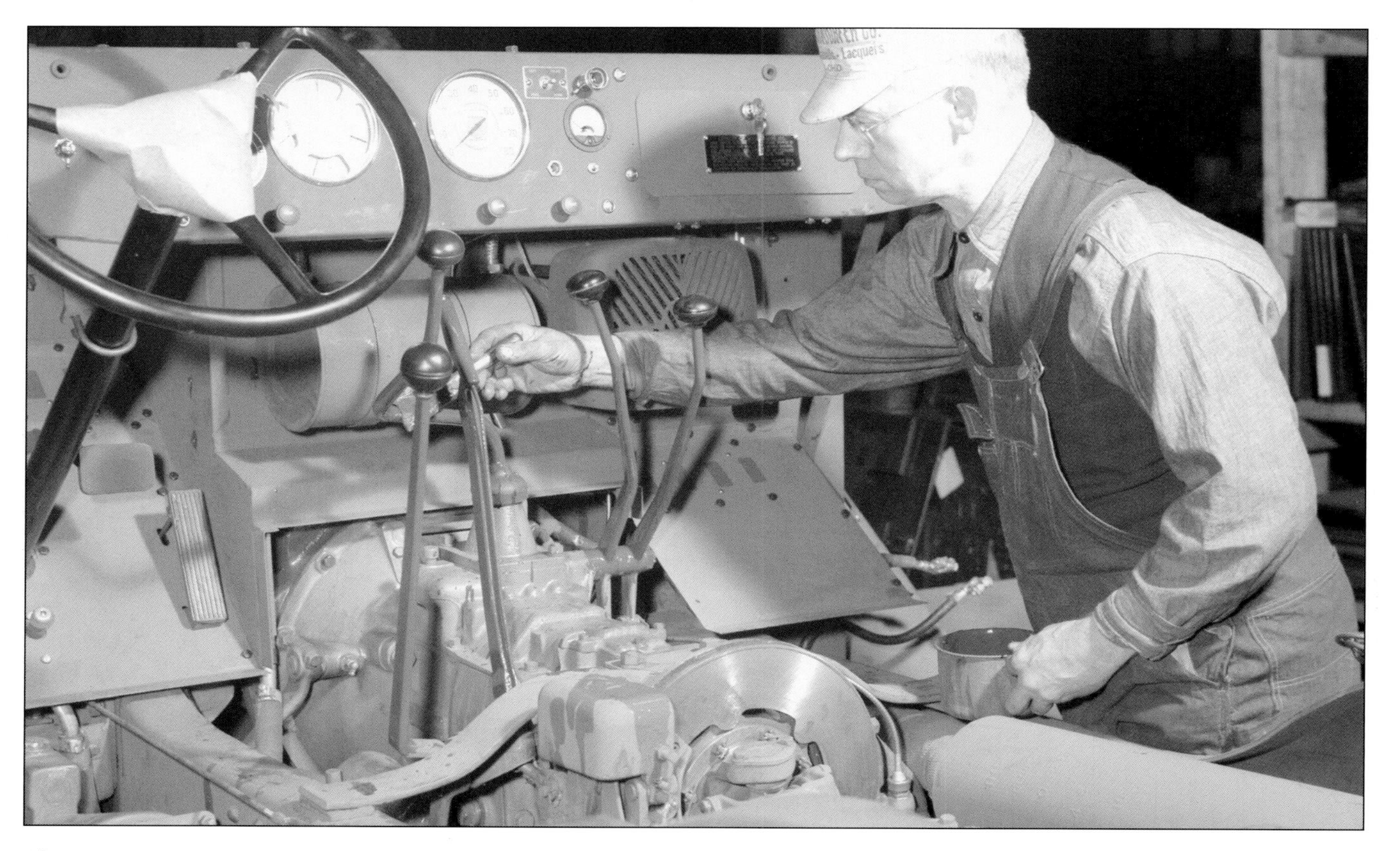

Above :*A worker at White Motor Company touches up the paint on a half-track's power takeoff shift lever. The other levers are, left to right, the transmission gear shift, transfer case shift, and front axle shift lever. To the left of the worker's paint can is the propeller-shaft brake disk. (LOC)*

Below: *This half-track chassis was photographed prior to installation of the body panels. To the rear of the jack shaft, as the rear drive axle was called, is a box-shaped, pressed-steel crossmember with three large lightening holes on top. This structure strengthens the chassis at the point where the bogies are attached. (LOC)*

Above: *A G-102 half-track cab is viewed from the right rear with the cab enclosure not installed. Underneath the dashboard are the radio filter (the horizontally positioned cylindrical object) and the cab heater. (NARA)*

Below: *A view of the left side of a half-track chassis illustrates the details present on the inner side of the chassis frame. The small, rectangular projections on the sides of the tracks were intended to assist traction. (NARA)*

Above: *The tracks fit tightly around the idlers, sprockets, and rollers, with virtually no sag visible on the top run. Note the exhaust to the front of the bogie bracket and the casting number on its rear face.* **Below:** *Three workers install the right-side track on a half-track chassis, while the man on the left installs a fender. To facilitate track installation, a chain hoist holds up the rear end of the vehicle, and shackles with turnbuckles support the suspension arms. At the bottom are tracks still in their shipping packaging. (NARA, both)*

Above: *An Aberdeen Proving Ground file photo illustrates the layout of the left side of the engine compartment with the hood and cowl removed. (NARA)*

Below: *As viewed from the right side of the half-track, to the front of the White 160AX engine is the oil filter; to the rear (left) is the air cleaner. (NARA)*

Above: *A test driver at the White Motor Company plant puts a half-track through its paces. The vehicle's body panels are not yet installed. (LOC)*

Below: *A half-track chassis undergoing testing has a canvas tarp lashed over the engine compartment. A makeshift windshield has been rigged in place. (LOC)*

Above: *A mechanic adjusts the carburetor of a half-track undergoing a test drive. The cylinder to the right, on the firewall, is the surge tank; next to it is the voltage regulator. (LOC)*

Below: *During a test run, a driver stops to jot down notes. Below the seat is the Bendix 371730 Hydrovac, used on later production vehicles. (LOC)*

Above: *During World War II, the Diebold Safe and Lock Company, Canton, Ohio, was converted to produce body parts for military vehicles, including half-tracks. Workmen at Diebold are installing 30-gallon non-leaking gasoline tanks on the chassis of a half-track.* **Below:** *A steel plate destined for a half-track body is dipped in a carbon compound during case-hardening at the Diebold Safe and Lock Company. (LOC, both)*

Diebold employees bolt sections of steel plate for a half-track body. (LOC)

Above: *A workman guides a partly finished half-track body as it is lowered onto the chassis at the Diebold plant. (LOC.)*

Below: *The rear panel of the body of a half-track is being installed. It is an M2-family vehicle, as indicated by the lack of a rear door .(LOC)*

Workmen assemble the rear panels on the right side of a half-track. (LOC)

Above: *A worker at Diebold installs the oil pan armor on the inner frame rail before the vehicle leaves the assembly line. The top of the radiator shroud is at the bottom. (LOC)*

Below: *A nearly complete half-track equipped with a winch has reached the end of the assembly line at Diebold. Note the early-production, civilian-type headlights. (LOC.)*

Above: *Partly completed M2 half-tracks proceed along the assembly line at the Diebold plant. Among other components, the fenders and radiator guards have yet to be installed. A skate rail appears to be present in the interior of the lead vehicle.* **Below:** *The row of M2s as viewed from the rear. The protruding structures at the lower rear of the body enclose the taillights and hold the bumperettes. They are closed in on the outer sides with solid triangular plates, a feature of Autocar M2s. White manufactured half-tracks feature open, triangular plates. (LOC, both)*

Above: *A view between rows of nearly completed M2 half-tracks at the Diebold Safe and Lock Company assembly line.* **Below:** *A mechanic checks the wheel of the M2 in the foreground while the mechanic on the next vehicle checks an engine component. The third vehicle still lacks its body components. (LOC, both)*

Above: *Ordnance Department inspectors check over M2 half-track cars at White's Cleveland plant before accepting the vehicles into U.S. service in December 1941. Each car received a bumper-to-bumper inspection. Here, they check the engine compartments of half-tracks with registration numbers W-4013384 and W-4013359. (LOC)*

Below: *Three half-track cars are readied for their final test-drive at the White plant in December 1941 before being delivered to the U.S. Army. The first and third vehicles have ammunition lockers to the rear of the side doors, indicating that these were M2 half-track cars or a derivative. (LOC)*

Above: *Half-tracks are parked at the White factory, complete except for the addition of Army-supplied equipment. The second vehicle, registration number W-4013362, is an M2. Details of the sides of the gear cases of the Tulsa winches are visible, as well as the sewn canvas covers for the winch drums. (LOC)*

Below: *What appears to be an Ohio state manufacturer's license plate is casually hung from the tow pintle of this M2 half-track that is passing a military police car with an Ohio license plate. (ATHS Archives)*

Above: *M2 half-track cars are nearing completion at the White Motor Company plant, Cleveland, Ohio, on 6 June 1941. Mechanics are tuning the engines. U.S. Army registration numbers in the W-4010898 to W-4010902 range are in view. (ATHS Archives)*

Below: *More M2 half-track cars approaching completion on the White Motor Company assembly line are shown in a photo taken on or around 6 June 1941. The third vehicle from the photographer bears U.S. Army registration number W-4010885. (NARA)*

Chapter 5
The M2 Half-track Car

Autocar and White Begin Production

Beyond the changes discussed in chapter 4, which were incorporated in the G-102 family of vehicles during production, other changes were specific to the M2, and later M2A1 vehicles, while still other changes were implemented to G-102 vehicles already in the field through changes incorporated in Modification Work Orders (MWO).

While the design of the M2 had been established, as with most combat equipment, further refinements were carried out during the course of production and field operations. OCM item 18257 of 13 July 1942 directed that track chains be installed in lieu of grousers to improve traction in mud and snow.

Also, in 1942 a surge tank was added to the cooling system to improve operation in high temperatures. The surge tank was also to be added to vehicles already in the field per Modification Work Order (MWO) G102-W8 of 31 August 1942. Coinciding roughly with this change was the adoption of a 2-pound CO_2 fire extinguisher in place of the 1-quart carbon tetrachloride unit previously used. This change was in accordance with OCM items 17204 and 18476 of 9 July 1942.

As far back as 29 April 1942 consideration was being given to installing heavier bogie suspension springs in light of the increasing weight of the vehicle. This change was approved, and on 17 March 1943 MWO G102-W23 was issued directing the change to be retrofitted to vehicles already in the field.

At about the same time, at White chassis serial number 234429, the radiator fan was changed from a six-blade unit to a five-blade model, which improved cooling. Changed at White serial number 250352 was the sediment bowl of the fuel filter. The original glass bowls were subject to cracking due to heat from the nearby exhaust manifold, so metal bowls began to be used instead. The other manufacturers implemented similar changes.

On 11 August 1942 the Office of the Chief of Ordnance issued a directive that a revised air cleaner be incorporated into production. This air cleaner would draw air from either the engine compartment or the driver's compartment. This was done in order to improve operation in dust, mud and cold weather.

Much more visible was the impact of a 21 August 1942 directive calling for the installation of mine racks on each side of the half-track body. This was in accordance with OCM 18585 of 6 August 1942. Modification Work Order G102-W21 issued 24 February 1943 called for the addition of these racks to vehicles already in the field.

Track throwing and suspension issues were at the forefront early in the production of the M2 and M3. On 4 September 1942 the Office, Chief of Ordnance issued a directive to provide a spring-loaded idler for the suspension, based on recommendations from Aberdeen Proving Ground. MWO G102-W14 was issued 29 November 1942 and covered a field modification developed by Field

Half-tracks leave the White factory in Cleveland, Ohio, in June 1941, on a demonstration drive. At least the first two vehicles are M2s. The small flaps near the centers of the tarpaulin covers are covering openings through which the antenna mounts could be raised. (NARA)

U.S.A.W-4010865

Service for use as an expedient pending quantity production of a permanent solution. On 15 July 1943 the latter became available in quantity, and MWO G102-W36 was issued describing the installation of the production-type spring-loaded idler.

As the half-track chassis began to be used for a variety of Gun Motor Carriages, new problems appeared. Among these was the breaking of headlights due to muzzle blast. In order to prevent this, on 7 November 1942 the Office, Chief of Ordnance directed that demountable headlights, comparable to those found on tanks, be incorporated into half-track production. While MWO G102-W34 of 21 June 1943 spelled out a field retrofit of these, it specified that the field modification was confined to Gun Motor Carriages, as prime movers and personnel carriers were not impacted by muzzle blast.

Less visible was a change initiated two weeks later, on 21 November 1942, when a directive, in response to numerous requests by the Signal Corps, was made to replace the radio suppression full shielding with resistor-bond type radio shielding.

Another fault exposed through field use had to do with troop transport. While negotiating rugged terrain, personnel in the back of the vehicles were frequently tossed about, potentially even out of the vehicle. Accordingly, hand grips were designed and incorporated into production and MWO G102-W22 of 25 February 1943 provided for the installation of these grips in vehicles already in the field.

White Service Bulletin 189, dated 10 March 1944, documented a much more visible change—the incorporation of luggage racks on the outside rear armor. The service bulletin documents that in compliance with Cleveland Ordnance District issue letter 62 these racks were incorporated as follows:

M2A1 with roller, contract W303-ORD-2080, White serial number 293090 to 293246 mixed, and all cars after 293249.

M2A1 with winch, contract W303-ORD-2080, White serial number 292592 to 292745 mixed, and all cars after 292746.

While U.S. forces utilized the bulk of the M2/M2A1 production, a total of 454 M2 half-tracks were delivered under Lend-Lease (excluding Theater transfers). Of those, 402 were to the USSR, 31 went to French Forces, eight to Brazil, and three to Mexico.

On 4 March 1948, OCM item 32058 reclassified the M2A1 as Limited Standard, and stipulated that it was "considered acceptable only for requirements other than Active United States Army."

On 10 February 1955 it was recommended that the M2A1 be declared obsolete. At that time the worldwide U.S. Army inventory of the vehicle was 57 units, with 1 unserviceable in Zone of Interior Depot Stock, 1 serviceable in Zone of Interior Depot stock, and 55 on hand in area depots.

Above: *The pilot M2 half-track car was photographed at Aberdeen Proving Ground, Maryland, on 9 April 1941. Protruding above the body are an M2HB .50-caliber machine gun and two M1917A1 liquid-cooled .30-caliber machine guns. (Ordnance Museum)*

Below: *M2s not equipped with a roller at the front of the frame were fitted with a Tulsa Model 18G winch with 10,000-pound capacity. A tarpaulin is installed over the cab and crew compartment. The upper panel of the door is folded down, and a curtain with a clear plastic window covers the opening. (ATHS)*

APG - 387

Above: *Ammunition lockers are built into the sides of the crew compartment of the M2 half-track, to the rear of the driver's and passenger's doors. These lockers have exterior doors, hinged at the bottoms. Inside the crew compartment, a door at the top of each locker permitted crewmen to access the top shelf of the locker without having to leave the vehicle. (ATHS)*

Below: *The M2 with tarpaulin installed is viewed from the left and right rear. The rear curtain of the tarpaulin had a flap at the center to allow access to the crew compartment. Protruding from the top center of the tarpaulin was a flap to accommodate a radio antenna when installed. (NARA)*

Above: *Tucked between diamond-tread steps and the bumperettes at the rear of the M2 are the taillight assemblies. Next to the left taillight is an external receptacle for trailer light and brake connections. On the rear of the body are straps and footman loops for stowing one M3 tripod for the .50-caliber machine gun and two M2 tripods for the .30-caliber machine guns.*

Below: *A small stowage box is fitted to the rear of the body of this early M2 half-track. The structure of the bumperette/step assemblies is visible; triangular gussets on the outsides of these assemblies served to strengthen them. Rising above the crew compartment is a pylon mount for a radio antenna. (ATHS, both)*

Above: *Early M2s had fixed (but adjustable) idler spindles that would not flex when the vehicle was operating on rough ground, which occasionally resulted in thrown tracks or damage to the suspension. This White-manufactured M2, registration number 4023630, shows the early-type fixed mounts for the idlers. Spring-loaded idlers were introduced to the M2 in September 1942. (ATHS)*

Right: *A half-track is undergoing assembly. The man to the right is wiring the left service headlight, while two mechanics work on the engine and its accessories. (NARA)*

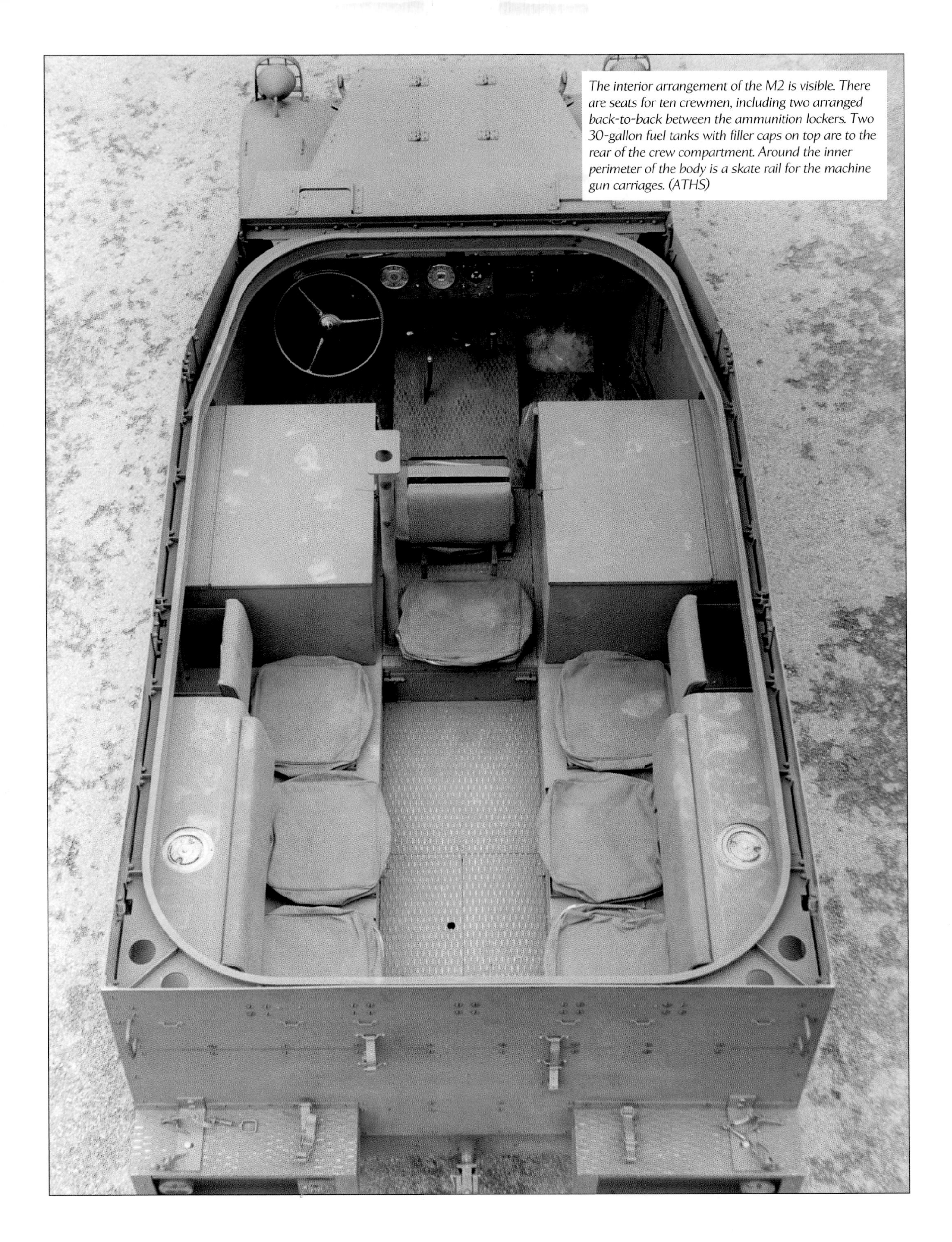

The interior arrangement of the M2 is visible. There are seats for ten crewmen, including two arranged back-to-back between the ammunition lockers. Two 30-gallon fuel tanks with filler caps on top are to the rear of the crew compartment. Around the inner perimeter of the body is a skate rail for the machine gun carriages. (ATHS)

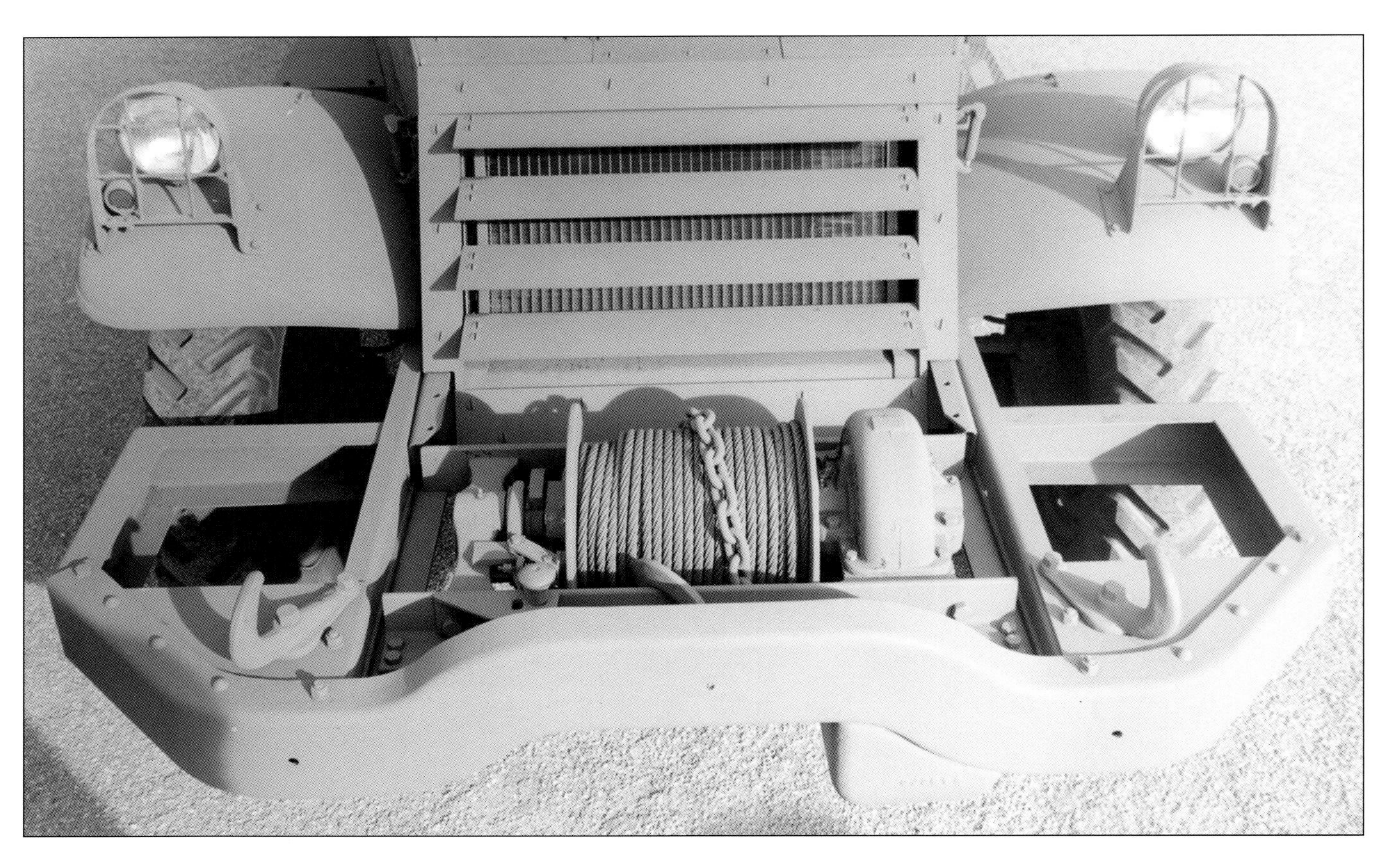

Above and below: *A close up view of the front bumper, Tulsa Model 18G winch, and early, non-dismountable headlights. Diagonally-oriented tow hooks are fastened to the gusset plate to the rear of the bumper. The armored louvers at the front of the body can be closed to protect the radiator from shrapnel and small arms fire. (ATHS)*

Above: *Although Tulsa Winch rated the 18G at 20,000 pounds, the military data plates list the capacity at 10,000 pounds. (NARA)* **Below:** *The layout of the instrument panel of the M2 half-track is similar to that of the M3-series half-tracks. The windshield comprises two plates of hardened, bullet-resistant glass in steel frame. Its protective armored shield is ½-inch thick and was usually raised when not in combat conditions. Above the windshield is the front section of the skate rail. (ATHS)*

Above: *The ammunition locker doors each have two latch plates bolted to them. When the doors are closed, these plates fit over two studs attached to the body over the door opening. The pins with the retainer chains can be inserted into a hole in each stud to hold the doors shut. Inside the locker are steel shelves for ammunition boxes. (ATHS)*

Below: *On the inside of the driver's and front passenger's doors are two telescoping rods that held the upper plate of the door rigid when raised. On the left side of the door are the latch and handle. Above the latch is a simple, sliding-bolt lock. Below the door are stowage brackets for a shovel and a mattock. (ATHS)*

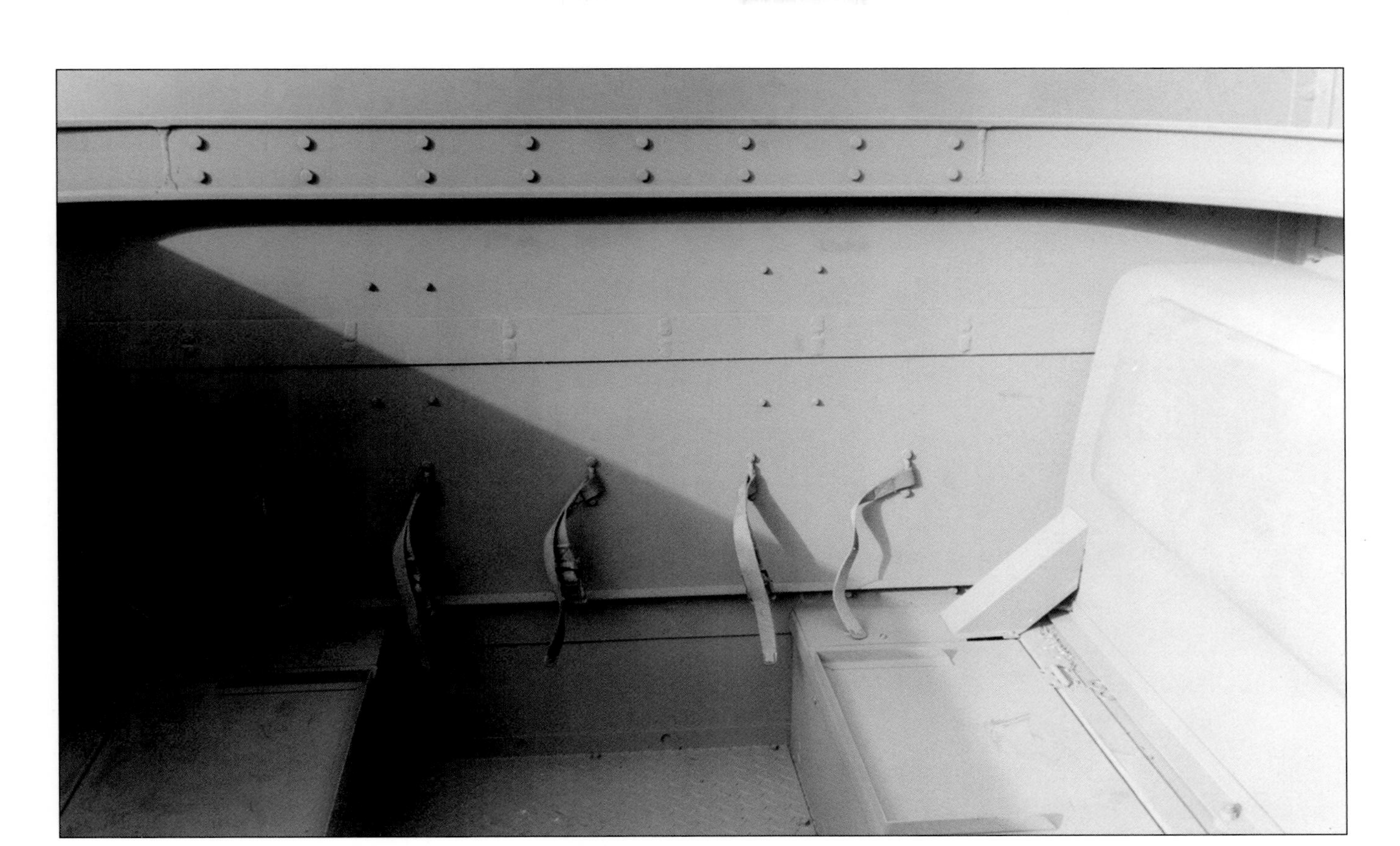

Above: *The side bench seats of the M2 are hinged, and when open they provide access to stowage bins below. Unlike the M3 personnel-carrier half-track, there is no door at the rear of the M2, so its crewmen had to enter and exit the vehicle over the sides. The webbing straps on the rear wall are for securing ammunition containers. (ATHS)*

Below: *The top of the battery box of an M2 half-track has been removed, revealing the top of the 12-volt battery. Some, if not all, of the early M2s had their registration numbers hand-painted on the sides of the engine compartments at the factory. The reference lines that the painter drew for these numbers are still visible. (ATHS)*

Above: *The pilot M2 half-track was photographed at Aberdeen Proving Ground on 9 April 1941. The M2 was designed as a scout/reconnaissance car and a prime mover for the M2 and M2A1 105mm howitzer. The ammunition lockers on its sides were intended to hold 105mm ammunition, but these compartments could be put to other uses when the vehicles were not towing artillery. (Patton Museum)*

Left: *The pilot M2 half-track mounts a Browning .50-caliber M2HB machine gun, flanked by two Browning M1917A1 liquid-cooled .30-caliber machine guns. The armored shield is lowered over the windshield, showing the two visors with sliding covers mounted in it. A steel tow rope is wrapped around the tow hooks. (Patton Museum)*

Above: *Shown in its intended role as an artillery prime mover, the M2 pilot vehicle has an M2 105mm howitzer hitched to its towing pintle at Aberdeen Proving Ground. The early M2s had directional tire treads. Later in production, these were changed to non-directional treads.* **Below:** *Although the pilot M2 is displayed with a 105mm howitzer hitched to it, this vehicle would also be put to use towing other artillery pieces, including the 75mm gun. Later, in order to improve the mobility and rapidity of deployment of those pieces, both of these weapons were mounted on board half-tracks.* (Patton Museum, both)

Above: *A right rear view shows the pilot M2 prepared to tow an M2 105mm howitzer at Aberdeen. The three tripods strapped to the rear of the half-track's body were for deploying the machine guns dismounted. A radio antenna has not been mounted to the antenna pylon protruding from the crew compartment. (Patton Museum)*

Below: *An electrical cable is connected to the receptacle next to the left taillight to provide power to the howitzer's Warner electric brakes. In the cab is an electric brake-load control, a rheostat device by which the driver could adjust the braking power of towed artillery and trailers according to their weight. (Patton Museum)*

This half-track with "M2" and "170" stenciled in white on its side also wears an Aberdeen Proving Ground identification plate, number 400. A command radio set has been installed in addition to the stock transmitter-receiver. This photo conveys the complex curves of the fenders and the unusual style of the directional treads on the tires. (Quartermaster Museum)

The same M2 half-track car portrayed on the preceding page is shown from a different angle during testing at Aberdeen Proving Ground, Maryland. The small license plate on the brush guard reads "APG-400." (NARA)

Right: *This official photo shows a carriage for the skate rail of an M2 half-track (upper left) and two types of machine gun cradle-and-pintle assemblies. At the top right is a D36960 or D54075 cradle-and-pintle assembly. The cradle and pintle at the lower right is the Model D40733 used with the M49 ring mount installed on the M2A1 and M3A1 half-tracks. (NARA)*

Below: *A close-up of a Browning M2HB .50-caliber machine gun on the skate rail of an M2 half-track. The gun is mounted on a D36960 cradle-and-pintle assembly with an elevating mechanism attached near the rear of the cradle. The ammunition tray on this particular type of gun mount has a beveled bottom. (NARA)*

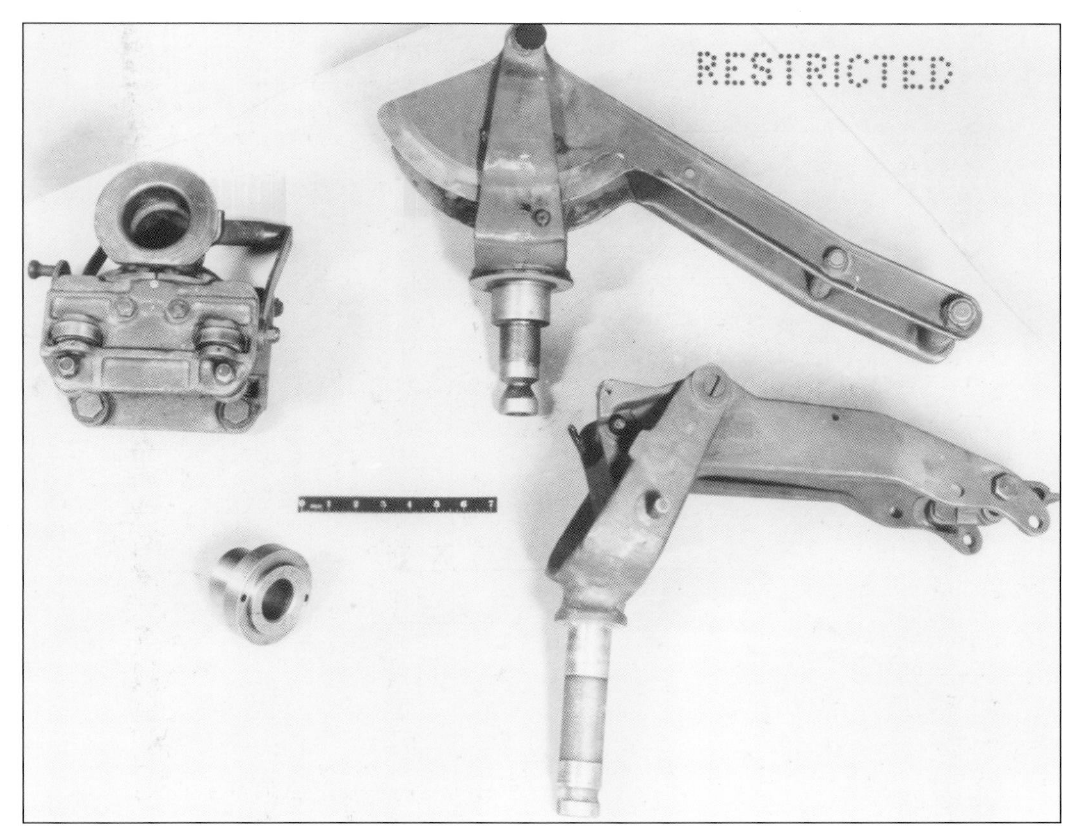

Above: *An overhead view of the receiver of a Browning M2HB .50-caliber machine gun on the skate rail of an M2 half-track. The number 1 is painted in white on top of the receiver and the ammunition tray cover. At the top are the front passenger's seat and the top of the right ammunition locker. (NARA)*

Below: *Details of the front right corner of the skate rail of an M2 half-track car are displayed. The small hand wheel with a knob underneath the carriage is the canting wheel, for adjusting the lateral level of the machine-gun carriage. (PAS)*

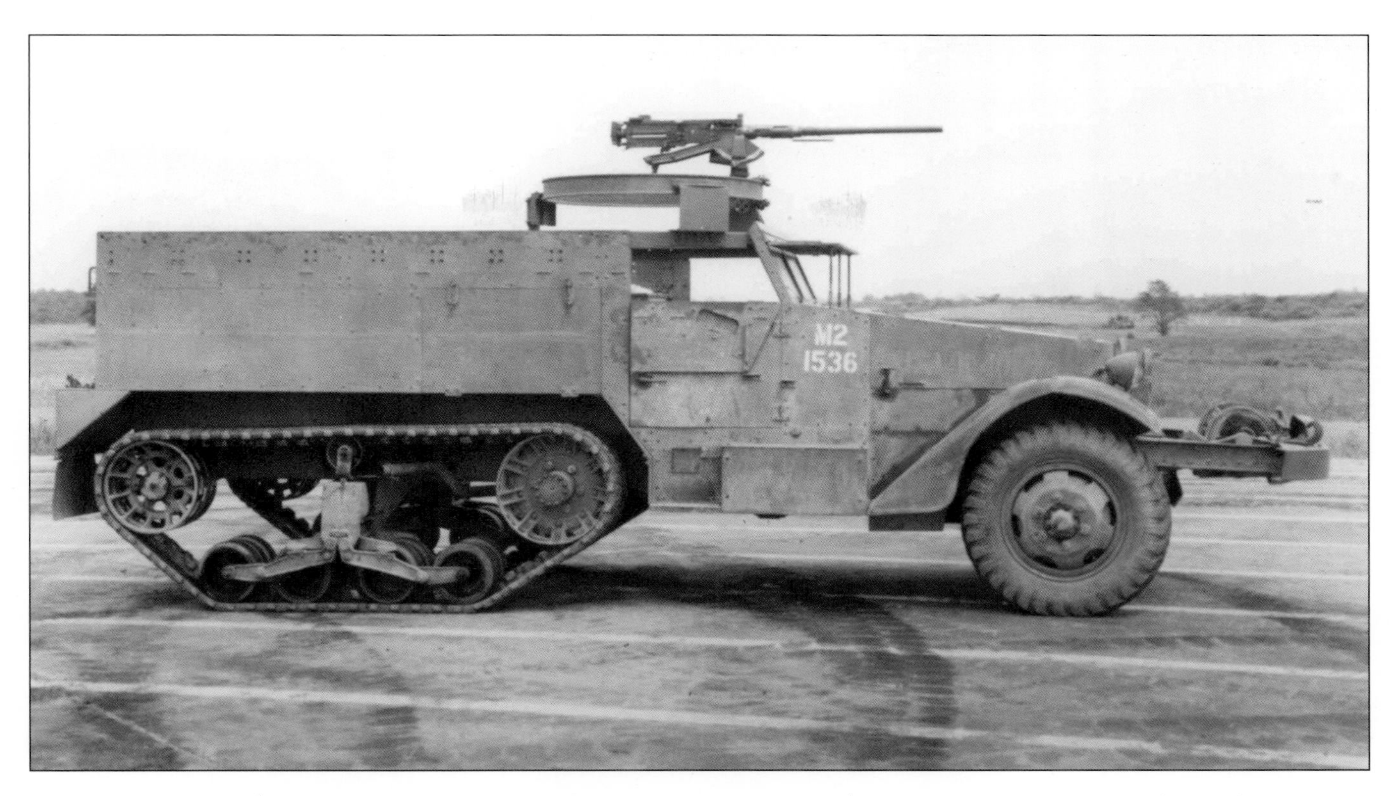

Above: *In May 1942, the Ordnance Committee recommended the removal of the skate rail in the M2 half-track cars in favor of an M32 ring mount over the front passenger's seat. Two M2s were modified in that manner and designated the M2E6. This is the first pilot, based on an early-production M2 modified at Aberdeen Proving Ground, where it was photographed on 3 August 1942. (NARA)*

Below: *As viewed from the front, the first M2E6 bears Aberdeen Proving Ground identification plate number 397 attached to the left brush guard. The M32 ring mount gave the gunner a higher operating position and allowed him to bring the gun into play at any angle around the vehicle much more quickly than with the skate rail, but at the expense of increased exposure to enemy fire. (NARA)*

Above: *On the second M2E6 half-track, the ring mount was attached to an armored superstructure covering the front and right side of the mount. This arrangement afforded the gunner better protection, and this pilot became the basis for a new model, the M2A1 half-track. The vehicle was photographed during trials on 1 February 1943. Note the .30-caliber machine gun on a pintle mount toward the rear. (TACOM LCMC History Office)*

Below: *Although this photo shows an M37A1 ring mount, this is the same ring, carriage, and cradle assembly that makes up the M49 ring mount installed on the standardized M2A1 half-track. Here, the Browning M2HB machine gun has been installed. Note how the mounting flanges are welded to the ring. (Rock Island Arsenal Museum)*

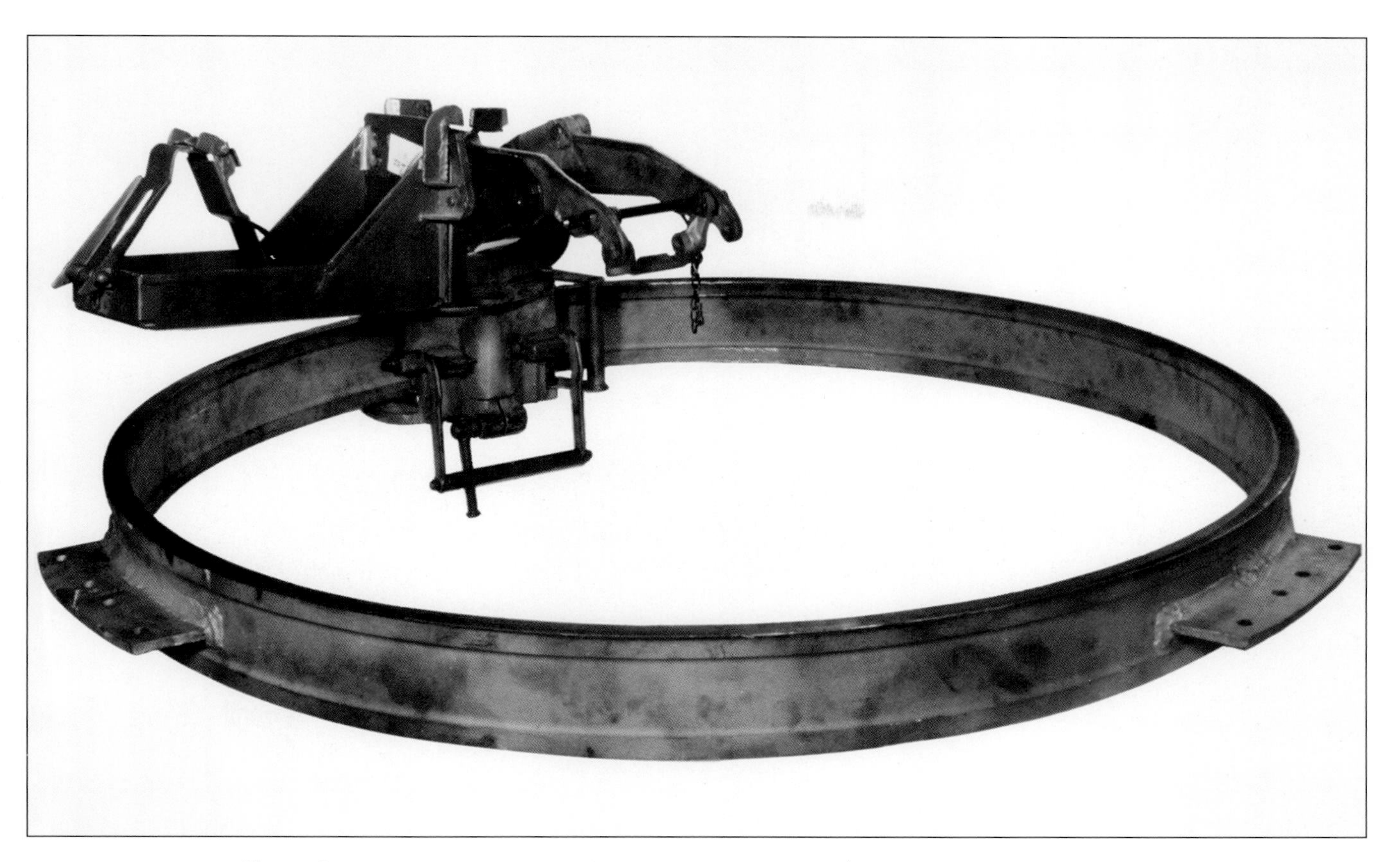

Above: *Shown with the machine gun removed, the elements of the mount are apparent, including the track and mounting flanges (Model C90771), cradle assembly (D40733), and carriage assembly (D40721). The ammunition box tray (D90078) is installed on the cradle. The horizontal bar below the carriage is its brake lever. (Rock Island Arsenal Museum)*

Below: *The first of two photos taken at Aberdeen Proving Ground in late January 1944 to document Project 6-11-29-1, a modification to mount an ammunition box support tray on the cradle of an M49 ring mount. Illustrated here is a .50-caliber ammunition box M2 on the tray. (NARA)*

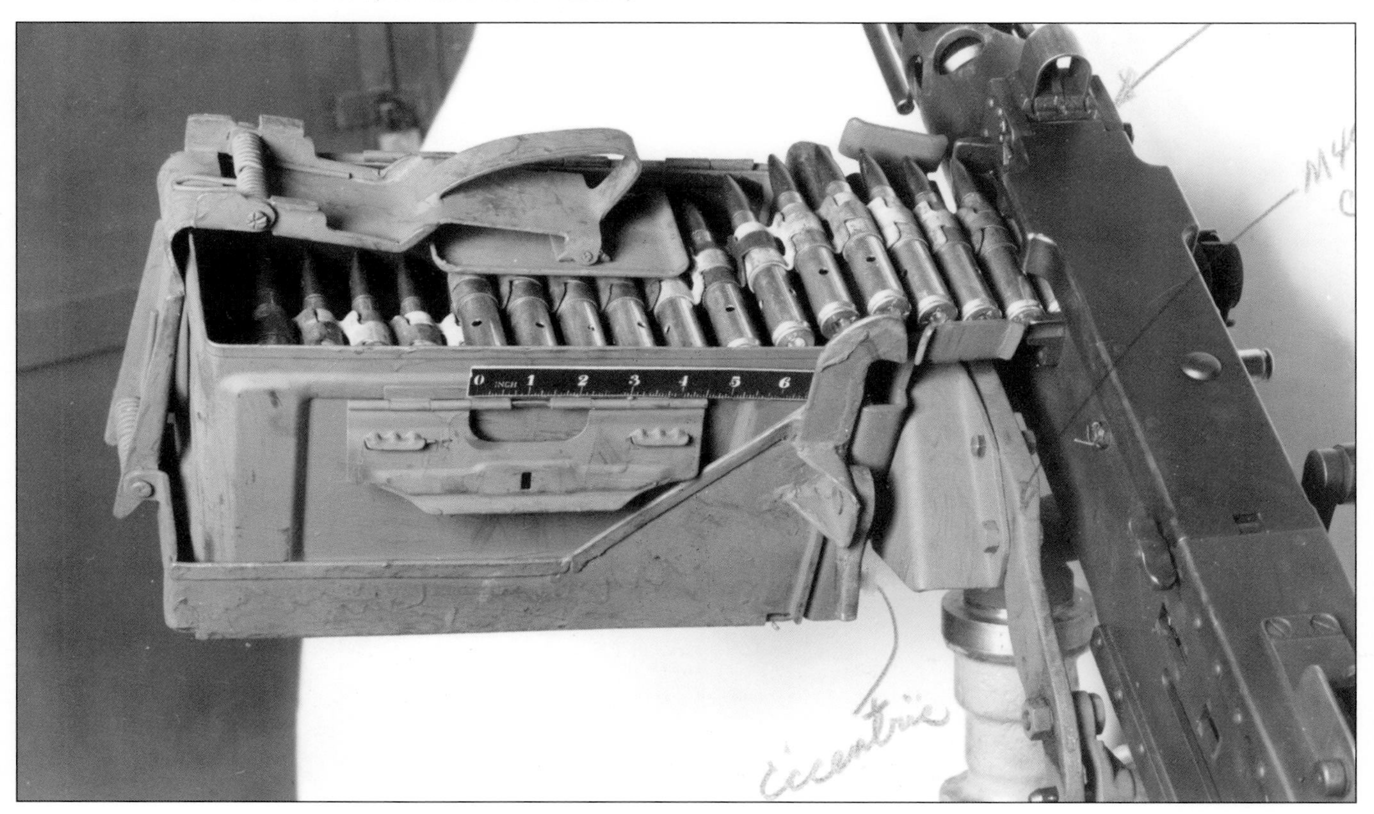

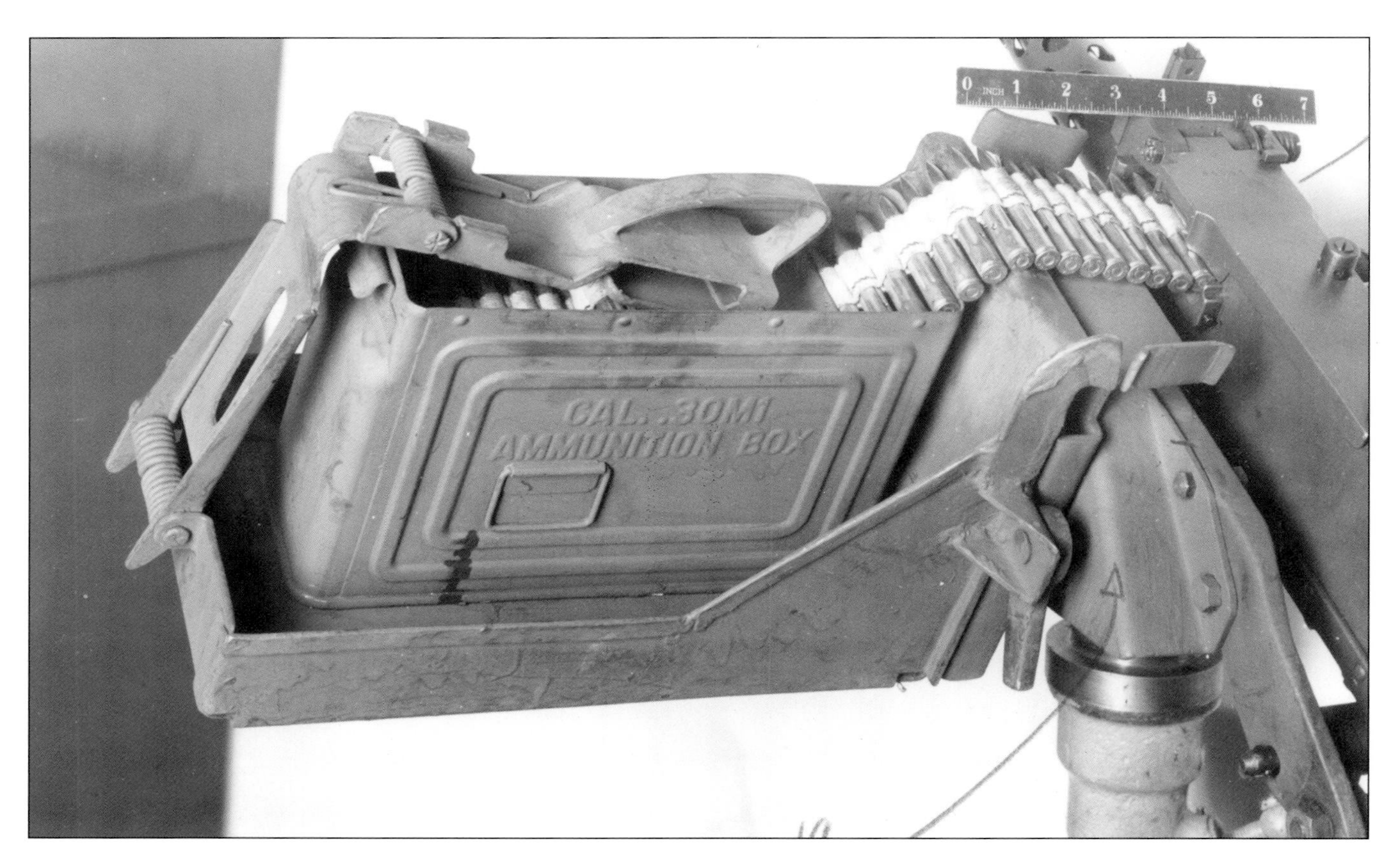

Above: *The second of two photos from APG's documentation of modifications to mount an ammunition box support tray on the cradle of an M49 ring mount. In this shot, a .30-caliber ammunition box M1 is mounted on the tray. Rulers are included for scale. (NARA)* **Below:** *Production of the M2A1 began in October 1943 and continued until March 1944, with a total production of 1,643. This late example was photographed at the Engineering Standards Vehicle Laboratory in Detroit on 15 February 1944. Its features include mine racks on both sides, 5-gallon liquid container racks, and detachable headlights with redesigned brush guards. (Patton Museum)*

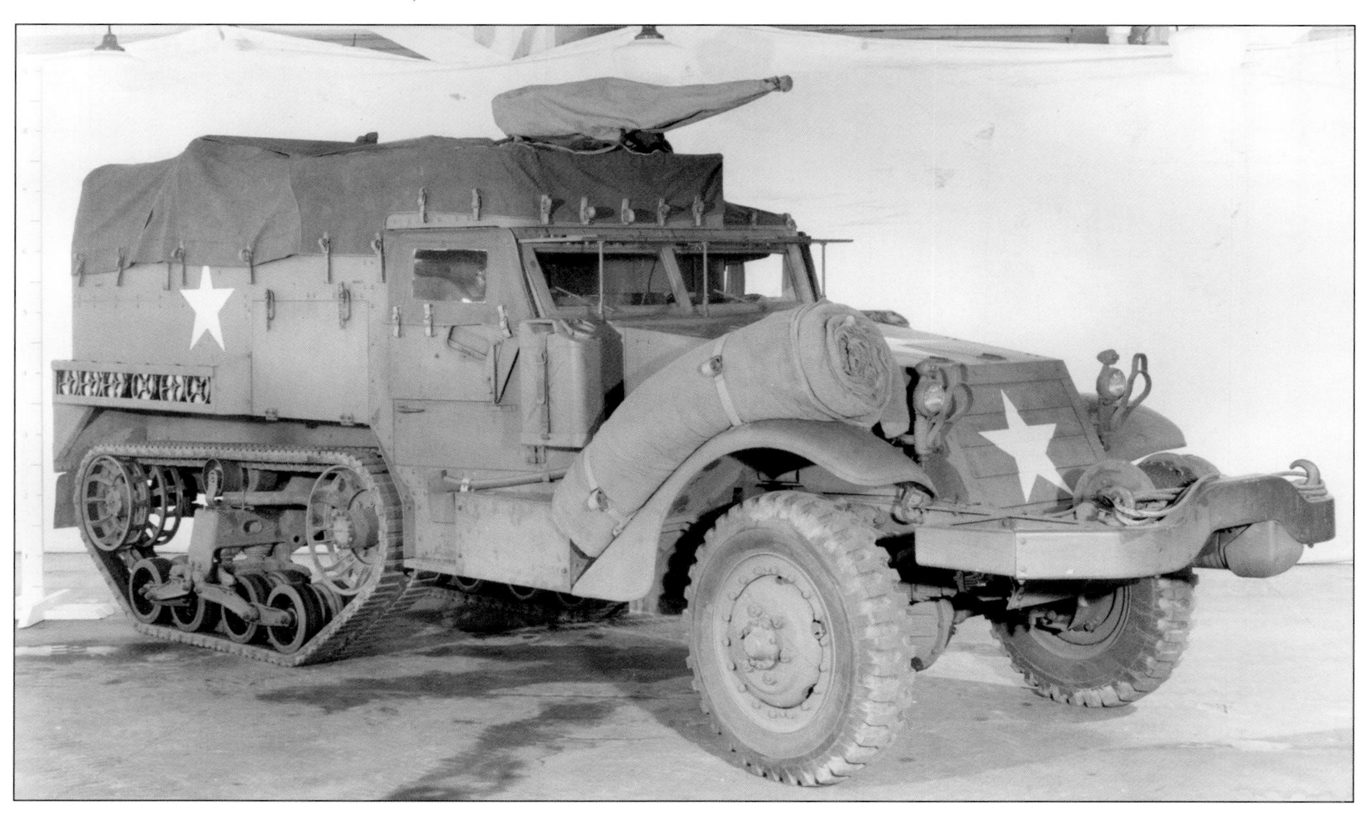

Above: *The tarpaulin, door curtains, and .50-caliber machine gun cover were part of the standard accessories. The tarpaulin is secured to the body with straps through footman loops. Bows support the tarpaulin. Protruding through the tarp to the rear of the cab is the radio antenna mount. (Patton Museum)* **Below:** *"Detroit Office, Chief of Ordnance" is stenciled in white on the ammunition locker door. Other markings include "M2A1 Half Track Car" on the side of the body, white national stars, and the number 49 in several places. The small stencils on the ammunition locker door read "SCR-528" and "Prepared by L.T.D. 11/27/43." SCR-528 is a reference to the type of radio set on board the vehicle. (Patton Museum)*

Above: *The same late M2A1, as viewed from the left side. In addition to the ring mount for the Browning M2HB .50-caliber machine gun, there are three sockets for pintle mounts for a single, moveable M1919A4 .30-caliber machine gun in the crew compartment. There is one mount is on each side of the body and one on the rear. (Patton Museum)* **Below:** *The M2A1 is fitted with the late-type idler-wheel springs. The operation manual for the vehicle specifies that the canvas tarpaulin cover for the cab and crew compartment be stowed on the right fender and a 12' x 12' canvas tarpaulin be stowed on the left fender. (Patton Museum)*

Above: *The Tulsa 10,000-pound winch is behind the center of the front bumper. Cast in small numbers on the front of the winch gear case is its part number, D-48226. The front axle with off-center differential is suspended on two elliptical leaf springs and shock absorbers. The detachable headlights are now mounted to the radiator side armor. (Patton Museum)*

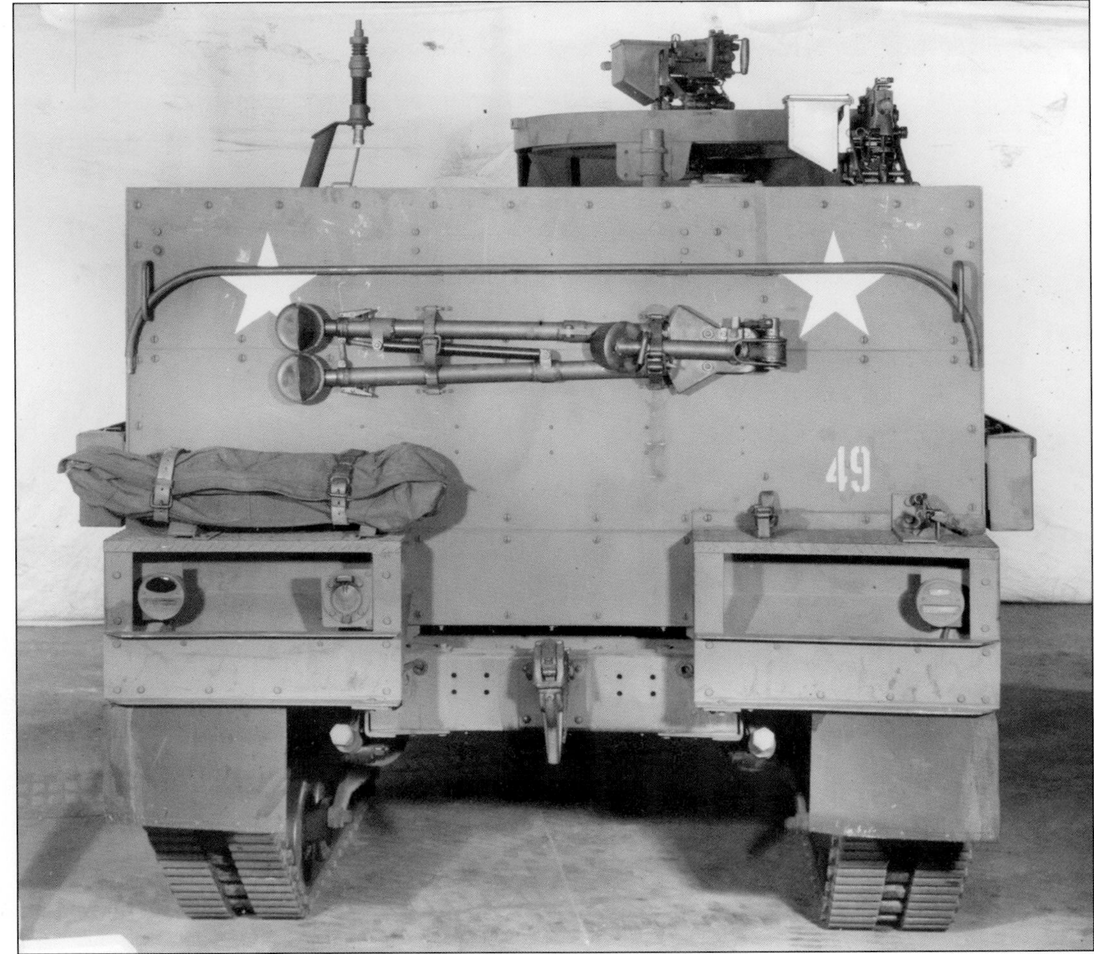

Below: *An M3 tripod for the Browning M2HB .50-caliber machine gun is strapped to the center of the rear of the body. Since only one M1919A4 .30-caliber machine gun was issued to each M2A1 half-track, there is one M2 tripod for that weapon. It is secured to the left step and is protected by a canvas cover. (Patton Museum)*

Above: *An overhead photo of a White M2A1 at the Engineering Standards Vehicle Laboratory, taken on 2 March 1944, reveals the shape of the ring mount and its armored supports and braces as well as that of the front bumper and its frame. The locations of the three pintle sockets for the .30-caliber machine gun are visible. The cradle is mounted on the right-side socket. (Patton Museum)*

Below: *Details of the inside of the right door of an M2A1. The framework at the right of the photo is a holder for a lubrication guide for the vehicle. To the left of the holder, shown in its retracted position, is one of two telescoping supports for the folding upper panel of the door. The design of the seats and their cushions is also visible. (Patton Museum)*

Above: *In a view of the rear of an M2A1 half-track from next to the ring mount (lower left), the rear and left-side pintle sockets are visible. Four boxes for .30-caliber ammunition are strapped to the back wall. One service headlight and two blackout headlights are also stowed on racks on that wall. (Patton Museum)* **Below:** *Here, the radio cabinet door is raised, exposing a faux SCR-508 made of wooden blocks during this July 1941 trial installation at APG. In the foreground is a MP-37 mast base. (NARA)*

Above: *With the right side of the hood open, that side of the White 160 AX 6-cylinder in-line engine and its components are in view. From left to right are the oil-bath air cleaner, air hose, carburetor, upper coolant hose, and FRAM oil filter (filter body no. 5267). To the right is the radiator. The curved oil filler pipe is next to the exhaust manifold. (Patton Museum)*

Below: *On the left side of the engine, the water pump is toward the upper front. The generator is partially visible below it. Below the distributor is the oil cooler. Toward the right are the surge tank and its pressure cap, part of the cooling system. The engine nomenclature and serial number are stamped on the crankcase next to the generator. (Patton Museum)*

Above: *A Colonel Grey poses next to the first M2 half-track car produced by White Motor Co. at that company's general offices in July 1941. The U.S.A. number W-4010737 is visible on the side of the hood, and Aberdeen Proving Ground license plate number 387 is on the bumper. (Tom Wolboldt collection)*

Below: *M2 half-track U.S. Army registration number 4011811 is loaded with civilian men and bears on the side the inscription, "U.S. ARMY / HALF-TRAC SCOUT CAR / Built By / THE WHITE MOTOR COMPANY / CLEVELAND." This livery was for promotional purposes, part of a demonstration for the public. (NARA)*

Above: *A Willys MA shows off its ability to climb stairs as the M2 half-track shown in the preceding photo and now loaded with even more men sits parked below. Demonstrations like this helped build local pride in companies producing materiel for the nation's military buildup for World War II. (NARA)*

Below: *An Army crew puts a White M2 half-track car through a test run at Fort Hayes, Ohio, on 6 June 1941. The chains on the front tires help this vehicle surmount the muddy embankment it is trying to negotiate. Notice the compression of the rear bogies. (NARA)*

Above: *The same M2 half-track car seen in the preceding photo climbs a steep bank at Fort Hayes, Ohio, on 6 June 1941. The M2 half-track was capable of climbing a 60 percent grade, depending on the condition of the surface of the slope. (NARA)*

Below: *Another M2 kicks up waves while driving in the same stream. No special precautions were necessary before or after operating for brief periods in water up to the maximum fording depth of 32 inches. (NARA)*

Above: *During Army tests of half-tracks, a front roller-equipped M2 drives down a stream. In the distance, another half-track is on the creek bank.* **Below:** *A group of M2 half-track cars proceed along a nearly dry creek bed during an exercise at Fort Hayes, Ohio, on 6 June 1941. Several of the cars are not armed, while many others display water-cooled .30-caliber machine guns of the M1917 family. (NARA, both)*

Above: *In a final photo from the 6 June 1941 army tests of White M2 half-tracks at Fort Hayes, Ohio, the vehicles negotiate a creek bed. Tests such as these allowed the army to determine the suitability of the M2 for service and to discover and correct any flaws. (NARA)* **Below:** *Six M2 half-track cars are lined up along a street. No evidence is visible even under close inspection of U.S. Army registration numbers, so these vehicles apparently were awaiting government inspection and acceptance. It is possible that these vehicles are convoying back to the White plant following armor installation by Diebold. (ATHS Archives)*

Even with production underway, evaluation of the vehicles continued in order to access potential design changes as well as to determine useful information to provide to the end users. The following series of photos shows the results of firing tests using various U.S. weapons. **Above and below:** *This is the result from hits made by .30-caliber ball ammunition. (NARA, both)*

Above: *Less successful was the test using .50-caliber ball ammunition. Considerable spalling has occurred along with several penetrations.* **Below:** *Paper has been taped to the interior surface of the doors to reveal penetrations by bullet fragments. At least two such penetrations can be seen around the visor. (NARA, both)*

Above: *A close-up of the opposite side of the vehicle shown above. The hits have caused some cracking along the fastening points.* **Below:** *The dramatic results of .50-caliber armor piercing and tracer rounds. This clearly illustrates the potency of the M2 .50-caliber machine gun—a weapon that found no equal among the U.S. Army's adversaries in WWII. (NARA, both)*

Above: *In the fall of 1941, testing also included these "grouser clamps" made by American Chain and Cable Co. This early attempt was not considered a success.* **Below:** *An Aberdeen Proving Ground photo dated 26 April 1941 documents the failure of the left bogie assembly after 102 miles of cross-country operation. Apparently the pin securing the rear outer suspension arm to the frame assembly sheared off. (NARA, both)*

Above: *Additional tests were conducted to determine the operating limits of the M2. Here, the vehicle has finally been stopped by approximately two-feet of Aberdeen Proving Ground mud. (NARA)*

Below: *This photo, taken at Pine Camp New York, illustrates the twin rear luggage racks which were originally furnished as part of a winterization kit. However, these racks were incorporated as standard equipment in early 1944. (NARA)*

Above: *Troops of the 8th Cavalry Regiment prepare to unload their M2 half-tracks at Anacoco, LA on 27 July 1941. They will be participating in 3rd Army maneuvers there. (NARA)*

Below: *Soldiers of the 3rd Army Engineers fire their M1917 .30-caliber water-cooled machine gun from the rear of an M2 half-track during night maneuvers in Louisiana on 23 August 1941. (NARA)*

Above: *An M2 half-track is parked at a rural store in Louisiana in the summer of 1941. No markings other than the U.S. Army registration number and a white stencil on the interior of the driver's door are visible.* **Below:** *In a photo taken near Camp Polk, Louisiana in September of 1941, M2 half-tracks cross a wooden pontoon bridge. The vehicles and guns belong to Battery B of the 78th Field Artillery Battalion. The towed weapon is a 75mm Gun M1897 A4. (NARA, both)*

Above: *Elements of Battery A of the 78th Field Artillery Battalion pause during the September 1941 maneuvers in Louisiana. When fully crewed for serving the 75mm gun, the M2 was a crowded place, with nine crewmembers onboard. (Office of History, U.S. Army Corps of Engineers)* **Below:** *During Third Army maneuvers in, Louisiana on 8 September 1941, an M2 half-track of the 66th Armored Regiment crosses a gully. An improvised stowage rack with two braces has been added to the rear of the vehicle along with a bracket to hold five 5-gallon liquid containers. Note the chalk tactical symbols. (NARA)*

Above: *An M2 half-track with a front winch is being used as a communications vehicle, probably during U.S. Army prewar maneuvers. A large wooden box is mounted on the rear of the body. Roughly applied i.d. markings are on the hood and the body. (Jim Gilmore collection)* **Below:** *Umpires confer with tank and infantry officers during Third Army maneuvers in Louisiana on 18 September 1941. A radioman in the M2 half-track in the foreground wears a pair of headphones over his garrison cap. Two Browning M1917 .30-caliber machine guns are mounted on the skate rail. (NARA)*

Above: *GIs ham it up for the cameraman during the September 1941 Louisiana maneuvers. This M2 is part of A Battery of the 27th Field Artillery Battalion. The 27th was part of the Red Army during the war games. Their large red banners can be seen in the background. (NARA)*

Below: *M2 half-track machine-gun crews hone their skills at a firing range at Camp Polk, LA while observers standing on the hoods look on. The crews of at least the first four half-tracks fire Browning .30-caliber air-cooled machine guns mounted on the skate rails. (NARA)*

Above: *In this shot, crews are trying their hand at the much larger .50-caliber weapon. On this range the targets have been raised and are contained within large wooden frames. Both photos were taken on 6 July 1943. (NARA)*

Below: *During the Tennessee maneuvers of June 1941, a 2nd Armored Division M2 half-track car fords a creek. Hitched to the rear of the half-track is a 37mm antitank gun. No markings are visible on the vehicle. (NARA)*

Above: *With at least nine crewmen on board, an M2 crosses a rickety, makeshift bridge over a creek. The skate rail is visible next to the driver's head. The outlines of the ammunition locker door to the rear of the driver's door are also faintly visible. (NARA)*

Below: *"TEXAS" with a thunderbolt running through it is painted above the storage locker door on this M2 half-track. A .50-caliber machine gun and a water-cooled .30-caliber machine gun are on the skate mount. (NARA)*

Above: *The crew of an M2 half-track poses with serious expressions on their faces for their photo during a road march, most likely during one of the Tennessee maneuvers.* **Below:** *As an M2 half-track ventures into heavy Tennessee brush in the fall of 1942, crewmen man the two M1917 .30-caliber, water-cooled machine guns. The unit code "8- 12E" faintly visible at the left probably indicates that this vehicle was assigned to the 12th Engineer Battalion, 8th Infantry Division. (NARA, both)*

Above: *The crew of an M2 half-track car pauses for a break during 10th Armored Division maneuvers in Tennessee in July 1943. A small U.S.A. number, 4014045, is on the hood, and the vehicle is equipped with a front winch, a machine-gun pedestal, and antenna mount. (Patton Museum)* **Below:** *M3 medium tanks and half-tracks of the 1st Armored Division gather during the start of autumn maneuvers in South Carolina. All of the vehicles are marked with prominent national stars. (PAS)*

Above: *An M2 of the 6th Infantry Regiment's anti-tank company gets a helping hand from an M3 tank while negotiating the sandy pine loam of South Carolina. The 6th Infantry was part of the 1st Armored Division during the maneuvers held there in the fall of 1941.* **Below:** *Red Army defenders secure a crossroads from their M2 half-track during the Carolina maneuvers in the fall of 1941. Their unique headgear appears in several shots of this force. The hats may have been designed to make their wearer appear more European. Also note the colored banner on the side of the vehicle's superstructure. (NARA, both)*

M2 Half-tracks of the 1st Armored Division approach the city of Rock Hill, South Carolina on 1 November 1941. The 1st AD was arriving by rail to participate in the Carolina maneuvers. (NARA)

As the trains arrive in Rock Hill, they immediately begin offloading their cargo. These vehicles were from the Headquarters Company of the 68th Armored Field Artillery Battalion. Note that several of these half-tracks have had their driving compartment canvas folded back. (NARA)

Above: *Unloading continues in Rock Hill. During this early pre-war period, 1st Armored Division triangular insignia was largely and colorfully painted on the engine hood and the rear of the body.* **Below:** *The crew of this M2 is using the relative camouflage of a large shade tree to conceal it as it secures the crossroads of a small South Carolina town. Exact locations were rarely cited in original captions. This half-track already sports its more familiar "combat rim" wheels. (NARA, both)*

Above: *During First Army maneuvers near Chester, South Carolina, on 11 November 1941, an M3 37mm antitank gun and M2 half-track of the 8th Division guard a highway. Although the M2 was often used as a prime mover for 105mm howitzers, it could tow smaller artillery pieces, including the 37mm antitank gun.* **Below:** *An M2 half-track crew of the 13th Armored Regiment adopts a stance for delivering antiaircraft fire to defend a bridge along Route 1 near Bethune, South Carolina, in the fall of 1941. While one crewman mans an M1917A1 .30-caliber machine gun, another one aims his Thompson submachine gun. One such .45-caliber gun was part of the defensive arms of each M2 half-track. (NARA, both)*

Above: *Much to the chagrin of its single passenger, this M2 is negotiating a steep embankment in South Carolina. This photo was taken during the second large maneuver held there in November 1941. (NARA)* **Below:** *Two half-tracks of the armored forces, including an M2 in the lead, proceed down a trail during maneuvers at Fort Benning, Georgia, in the spring of 1942. The two soldiers visible in the front vehicle are wearing M1917 helmets left over from World War I. (LOC)*

Above: *Armored troops were photographed aboard M2 half-tracks during spring 1942 maneuvers at Fort Benning. For camouflage purposes, the white recognition stars have been overpainted with olive drab of a different tone than the vehicle's color. (LOC)*

Below: *An M2 half-track car, registration number W-4011345, is parked among several M3 half-tracks at Fort Benning in the spring of 1942. It is equipped with one .50-caliber M2HB machine gun and two M1917A1 .30-caliber machine guns. (LOC)*

Above: *G.I.s wearing garrison caps await transport in a fleet of half-tracks, including numerous M2s, at Fort Benning in 1942. The vehicles are notably devoid of the extra stowed equipment they would usually carry during overseas combat operations. At the bottom center is a radio antenna mount.* **Below:** *Although this photograph was originally used as a public-relations tool to explain why the military had priority in rail transportation, it provides a good view of M2 half-tracks being transported cross-country on flatcars. This was a common sight during World War II. (LOC, both)*

Above: *A wrecker has hoisted the left side of an M2 half-track to allow mechanics to perform repairs or maintenance on its underside. Note the left fuel tank drain plug on the underside of the body above the track.* **Below:** *The crew of an M2 half-track of the 12th Armored Division mans a position in a forest during field maneuvers. One of the men has his hand on the grip of a M1917 .30-caliber water-cooled machine gun. The U.S.A. number, W-4023366, is faintly visible. (NARA, both)*

Above: *An M2 half-track laden with troops and gear follows a Jeep along a trail next to a cornfield. The rack for five-gallon liquid containers on the rear of the half-track is a right-hand mine rack from an M2. The presence of this mine rack—in spite of its non-standard location—probably dates this photo to the fall of 1942. (NARA)* **Below:** *Pausing along a road, crewmen on an M2 half-track observe the terrain to the side, ready to bring the right M1917A1 .30-caliber machine gun into play. This vehicle has a stowage rack on the rear of the body, piled high with 5-gallon liquid containers, bedrolls, and other gear. The machine gun tripods have been moved from the rear of the hull to the left fender. (NARA)*

Above: *This M2 half-track, photographed at Fort Knox, Kentucky, has mine racks to the rear of the ammunition locker doors. These racks were added following a directive of the Office of the Chief of Ordnance on 21 August 1942, and they were retrofitted to many M2s already in service. "Falcon" is stenciled on the door, and "AFS" is painted in the center of the recognition star. (Patton Museum)*

Below: *This early production M2 still mounts a pair of water-cooled Browning M1917A1 .30 caliber machine guns. Note the early, cage style light guards and the directional tires. This photo also shows the interior of the side stowage box to good advantage. This crew of this vehicle has added an additional tool rack to the side of the superstructure. (Patton Museum)*

Above: *A column of three half-tracks has halted, and a crewman in each vehicle is making hand signals. The lead vehicle is an M2. The second in line (and probably the third) is a T19 105mm howitzer motor carriage. A canvas dust cover is fitted over the howitzer. Note the diamond unit symbol on the bumper and the nickname "Hell's Belle" on the door of the first half-track.* **Below:** *An M2 half-track comes over a rise during tests at Aberdeen Proving Ground. An APG identification plate is on the left bumper. This angle offers a good view of the underside of the front end of an M2. Note the protective sleeve over part of the barrel of the .50-caliber machine gun. (NARA, both)*

Above: *Half-tracks such as the M2 generally performed well in muddy terrain because their tracks contributed to relatively low ground pressure, keeping the vehicles from sinking excessively into soft ground. The armored louvers are open, revealing a glimpse of their operating linkages.* **Below:** *This M2 half-track was photographed at Aberdeen Proving Ground, Maryland, during the annual visit of a class of U.S. Military Academy cadets. An M2HB .50-caliber machine gun and M1919A4 .30-caliber machine gun are mounted on the skate rail, with protective sleeves fitted over their barrels. (NARA, both)*

Above: *This half-track has been secured to a flatcar "by the book." Wooden chocks have been pushed up to the tracks, a 2x4 cleat is attached to the car's planking next to the track, and tightly twisted baling wire secures the bogie frame to stake pockets on the side of this flatcar. The location is Camp Chaffee, Arkansas. (NARA)*

Below: *In a posed photo, the crew of an M2 half-track presents a strong front, with one G.I. manning the ring-mounted .50-caliber machine gun and another aiming a Thompson submachine gun. This half-track and its personnel probably belong to the 10th Cavalry Regiment during its service at Ft. Leavenworth, Kansas in 1941. (NARA)*

Above: *The crews of M2 half-tracks submit to inspection on 12 November 1942. They are members of the 93rd Armored Field Artillery, stationed at the time at Fort Chaffee, Arkansas. The T19 Half-track Motor Carriages of the unit can be seen in the background.* **Below:** *An M2 half-track of the 8th Cavalry moves past a civilian Piper Cub. This photograph was taken during 1st Cavalry Division maneuvers held at Ft. Bliss, Texas in July of 1941. (NARA, both)*

Above: *During a training exercise, a crewman aims the M1919A4 .30-caliber machine gun, with its carriage at the front center of the skate rail, while another man communicates over the radio. A finger-jointed wooden ammunition box holding a dummy ammo belt is attached to the gun cradle. Next to the barrel is an antenna on a pylon mount with the radio set adjacent to it.* **Below:** *On this M2 half-track, a light-colored square and "7C" are painted on the right side of the bumper, and a rectangle and "7C 30" are marked on the side of the crew compartment. (NARA, both)*

Above: *This M2 was photographed during exercises at the Desert Training Center. Special precautions were required when operating the M2 and M3 half-tracks and their derivatives in sandy desert conditions for protracted periods, such as lowering the air pressure of the tires for optimum traction and fitting cotton bags over the air cleaner and hydrovac cylinder to keep out excessive dust. (NARA)* **Below:** *Speeding down a sandy trail, an M2 half-track churns up considerable dust. In addition to the number 33 painted in white on the bumper, "AWOL" is painted on the bottom of the passenger's door. The crew has taken advantage of the raised armored windshield by using it for a stowage rack for a large roll, possibly a tent. (NARA)*

Above: *A heavy wrecker is lifting the front end of M2 half-track 4012352. This probably was part of a training exercise, as the M2 has been shorn of much of its on-vehicle equipment, such as the machine guns, pioneer tools, machine-gun tripods, and such. (NARA)*

Below: *M2 half-track 4010972 pauses during desert maneuvers. The crew has discarded their steel helmets in favor of the lighter weight, and undoubtedly cooler, denim utility caps. The pattern of dust accumulation on the rear armor is of interest. (NARA)*

Above: *The crew of an M2 half-track undergoes training maneuvers in the desert. The man standing in the vehicle is holding a radio handset. A .30-caliber machine gun is on the skate mount. Helmet liners, rather than steel pots, are the preferred headgear. (NARA)*

Below: *An M2 half-track passes a CCKW 352 short wheelbase truck. Given the depth of the tires in the photo, the truck may be having difficulty in the soft sand. The M2 is pulling 105mm M2 howitzer on the carriage M2A3. (NARA)*

Above: *An M2 half-track car proceeds along a trail at the Desert Training Center in California. At this base, the U.S. Army trained many of its units that were destined for fighting in North Africa, helping them learn the skills necessary for desert combat. (NARA)*

Below: *An M2 half-track precedes a 75mm gun motor carriage M3, during training in the UK in the fall of 1942. As an antidote to the balmy British fall weather, the crewmen are clad in tanker jackets and winter combat helmets, and some wear goggles. (NARA)*

An M2 half-track gently settles to the dockside of Reykjavik harbor on 4 March 1942. This equipment was part of a U.S. force sent there to relieve British troops who had occupied Iceland to prevent it from being seized by the Germans. In spite of these occupation forces, Iceland remained officially neutral throughout the war. (NARA)

Above: *This M2 "ONE DOZEN ROSES," has just come ashore at Les Andalouses, Algeria on 9 November 1942. It is towing an M3 37mm anti-tank gun and the full deep-water wading kit is still in place. Fifty caliber tripods are strapped to each fender.* **Below:** *An M2 and crew of D Co, 13th Armored Regiment, 1st Armored Division is seen at Souk el Khemis Tunisia on the morning of November 26, 1942. Note the large tent and the camouflage net on the ground. (NARA, both)*

Above: *Note the unit identification markings on the back of the luggage at the left rear corner and the left hand drive warning painted on the back plate of the M2. The Naval Officer (in the leather jacket) is Chief Petty Officer Jack Pennick, who served with Commander John Ford in the Field Photographic Unit.*

Below: *This early-production M2 half-track in North Africa was hit by a German 88mm shell just behind the right door on 21 December 1942. The vehicle, registration number 4011146, has the name "The Clipper" painted on the driver's door and an American flag on the side of the body. (NARA, both)*

Right: *Ordinance workers conduct maintenance on the rear suspension of the M2 half-track. The central suspension unit has been removed so that the various bearings and joints could be properly lubricated. (NARA)*

Below: *Due to the cramped interior of the M2, most vehicles had large stowage containers across the rear of the vehicle. In many cases, these were constructed unit wide, resulting in several interesting types. Here, an ordnance worker is welding perforated stock to the rear hull. Most likely, these perforations will be use to attach a wire mesh to retain gear. This rack is designed to clear the rear taillights and also to still facilitate the mounting of the .50 caliber tripod. Note the bows for the canvas top stowed through the grab handles. (NARA)*

Above and below: *This M2 was photographed in Oran in early 1943 as it was prepped for combat. Tripods for all of the skate rail-mounted weapons were carried on the rear of the hull. The larger .50-caliber tripod is seen here. The tripods for the two .30-caliber weapons were normally stowed beneath it on either bumperette, but the clasps are empty on this vehicle. (NARA)*

Right: *This rather well known photo depicts an M2 of the 601st Tank Battalion near El Guettar, Tunisia on March 23rd, 1943. In this instance the M2 is being used as a command vehicle. It contains at least one additional radio set as evidenced by the second antenna protruding from the crew compartment. An improvised mine rack has been added to the left side of the hull and a .50-caliber ammunition box has been fastened to the left side of the engine compartment. (NARA)*

Below: *Members of the 15th Engineer Battalion, 9th Infantry Division, take a break in Bizerte, Tunisia on 9 May 1943. Small U.S. flags are affixed to the headlight brush guards. Two machine-gun tripods are stowed on the left fender. (NARA)*

Above: *An M2 half-track and an M1 wrecker get a fresh coat of paint at the paint department of the Ordnance Base Shop in Casablanca, Morocco. Ordnance base shops handled the theater's heavy maintenance needs. This photo was taken on 28 July 1943. (NARA)* **Below:** *This M2 half-track has found new owners in the form of the German 501st Heavy Tank Battalion in Tunisia. Both the half-track and trailer are full of Tiger I road wheels, indicating that it is being used by the maintenance company of the battalion. Although not visible in this photo, other shots show that captured U.S. vehicles were elaborately marked with this unit's tactical symbols. (BA 557-1018-28A)*

Above: *This M2 is being used in the recon role while serving in Italy in the early winter of 1943. This crew has made good use of their fabricated rear stowage rack. Unlike their German counterparts, U.S. half-tracks had driven front axles. The tire chains, shown here, not only helped maintain steering control, but also aided in slogging through the snow, ice and mud.* **Below:** *M2 half-tracks of an engineering unit line up in preparation for loading in Torcross, England on 7 May 1944. With less than a month to go for the cross-Channel invasion, they are participating in practice landings. (NARA, both)*

Above: *The follow-up wave: an M2A1 sits on the deck of an LST as it travels across the English Channel on 8 June 1944. The hood of the vehicle is marked with the large "invasion" star partially composed of gas-sensitive paint. (NARA)*

Below: *"DIRTY GERTIE," an M2A1 pulling an M1 57mm gun, rushes past French civilians on the way to the battle of Cherbourg on June 23, 1944. An M4A1 DD tanks sits in the background. A staff car is among the convoy. (NARA)*

Above: *Reinforcements move up in half-track vehicles along a Normandy road to new positions west of St. Lo, where American troops broke through the German lines in the offensive that opened on 25 July 1944. (Image Bank WW2)* **Below:** *A U.S. Army half-track with "ZEPHYER" stenciled at the bottom of the door and an M5A1 light tank stand by at a crossroads as German prisoners of war are marched past. The scene is Sartilly, France, on 31 July 1944. Censors' marks obliterated the unit markings on the half-track's bumper, but there is a small placard on the right headlight brush guard that reads "HQ." (NARA)*

Above: *Civilians, somewhere in France or the Low Countries judging from their garb, are gathering to watch as several U.S. Army M2A1 half-tracks pass through their town. The unit markings on the bumpers have been censored, but the name "DARING" is visible on the bottom of the driver's door on the nearest half-track.* **Below:** *Children of Soignies, Belgium, come forward to greet members of the 41st Armored Infantry Regiment, 2nd Armored Division, in an M2 half-track on 12 September 1944. The name "ALABAMA" is painted on the bottom of the driver's door, and a large stowage bin filled with bedrolls, camouflage netting, and other gear has been installed on the rear of the vehicle. (NARA, both)*

Above: *An M2 half-track of the 2nd Armored Division crosses a temporary bridge near Maastricht, Holland in the fall of 1944. (Image Bank WW2)* **Below:** *Artillerymen manhandle a 57mm antitank gun towed by an M2A1 half-track in Aachen on October 15, 1944. Supplementary armor sections for the gun have been positioned on the sides of the crew compartment, perhaps more useful in providing protection for the half-track in the street-fighting environment. (NARA)*

Above: *An M2 half-track moves up a rise in the Hürtgen Forest on 26 November 1944. The vehicle is part of the 4th Cavalry and is being brought forward to help hold the terrain, which had been taken from the Germans the previous week. The other vehicles are part of a graves registration unit, which is recovering bodies from the earlier battles. Armor sections for the 57mm gun are also present here.* **Below:** *A soldier inspects the inside of the cab of a half-track knocked out by a German mine in Bining, France, in early December 1944. The vehicle, U.S.A. number 4025260, was assigned to the 495th Field Artillery, 12th Armored Division. Marked on the five-gallon liquid container is "V-80." (NARA, both)*

Above: *With extensive damage to the front wheel and axle, it's time to start looking for another mount. One GI begins the not insignificant task of emptying out the considerable amount of ammunition and gear from the half-track.* **Below:** *In Xosse, France an M2A1 half-track advances on 18 November 1944. A whip antenna is at the rear of the crew compartment. A fitted canvas cover is over the ring-mounted machine gun and the top of the pulpit. (NARA, both)*

Below: *This radio-equipped M2A1 is being employed as an infantry battalion command post a few yards inside the German border on 21 December 1944. A freshly painted unit symbol, known to be associated with the 12th Armored Division, is on the stowage compartment door.* **Below:** *A column of vehicles led by a winch-equipped M2A1 passes through a Belgian town during December 1944. Two crewmen stand in the ring mount. The snow-covered tarpaulin protects the rest of the cab and crew compartment. A stowage rack has been installed on the armored shield of the windshield. (NARA, both)*

Above: *A winch-equipped M2 half-track full of happy-looking G.I.s of the 1st Infantry Division rolls through Essentho, Germany, during the drive toward Buren in April 1945. It is towing an M1 57mm antitank gun. Supplemental shields can be seen stowed.* **Below:** *An M2A1 half-track plows its way down a muddy road. The vehicle belonged to the 16th Infantry Regiment, 1st Infantry Division. This photograph was taken on 15 February 1945 in the Hürtgen Forest. (NARA, both)*

Above: *Some of the original markings on this M2A1, photographed near Thionville, France, are faintly visible through the whitewash, including "HQ 65" on the left bumper. The "S" suffix in the registration number signifies that the vehicle passed a radio-interference suppression test. (NARA)*

Below: *The same M2A1 stands at the ready on 12 January 1945. The vehicle has a coat of whitewash for camouflage, leaving exposed a few thin strips of the original olive drab and the vehicle's registration number. A stowage rack has been added to the rear of the body. (NARA)*

Left: *A G.I. puts the finishing touches on the insignia artwork on the ammunition locker door of an M2 half-track. He evidently has wiped the area and pasted on the insignia, pre-painted on fabric or heavy paper—an easier and neater proposition than painting the insignia directly on the half-track. Traces of whitewash are visible on the vehicle. (NARA)*

Below: *In Vettweiss, Germany, in late February 1945, a U.S. Army M2 half-track with a 57mm antitank gun hitched to it pauses as infantrymen in the background look on. The name "PUBLIC ZOO" is painted on the door of the stowage bin to the rear of the driver's door. A dead German soldier lies in the foreground. (NARA)*

Above: *Several half-tracks of Combat Command B, 10th Armored Division, Third Army, are in the recently captured town of Irsch, Germany, on 26 February 1945. The half-track to the left has large stowage provisions on the rear, wrapped in canvas apparently for weatherproofing.* **Below:** *An M2 half-track leads a column through the town of Kestert, Germany in late March of 1945. Kestert sits along the Rhine (seen to the left) between Bonn and Frankfurt. (NARA, both)*

Above: *Soldiers are being transported on half-tracks of the 735th Tank Battalion, 87th Infantry Division in Saalfeld, Germany, in April 1944. The closest half-track has a large number 3 painted on the side.* **Below:** *A half-track is transporting German prisoners of war rounded up between Erboldshausen and Kalefeld, Germany, in April 1945. One of the POWs has his hand on the .30-caliber machine gun, undoubtedly unloaded, mounted on the side of the crew compartment. (NARA, both)*

Above: *A heavily laden M2 half-track car of the 13th Armored Division approaches an engineer bridge spanning a river between Germany and Austria on 6 May 1945. Contraptions have been added to the vehicle to carry equipment, such as the board fastened to the side above the track.* **Below:** *During the push into Germany in early 1945, a French M2A1 half-track car passes road signs for Nürnberg, Zeitz, and Leipzig. The vehicle is equipped with a large stowage box on the rear of the body and is towing a captured German 75mm PaK 40 antitank gun. (Heslop Collection, Brigham Young University, both)*

Residents of Pilsen, Czechoslovakia, serve beer to troops of the Third U.S. Army in early May 1945. The name "JOYCE ANN" is painted at the bottom of the door of the half-track. The object atop the netting on the armored windshield is a civilian hat. (NARA)

Left: *Members of the 37th Ordnance Battalion give an M2 transmission a make-over on 30 November 1942. The battalion was part of the 41st Infantry Division at this time and was stationed in Rockhampton, Australia. (NARA)*

Below: *On 26 December 1943, a half-track with an M1897A4 75mm gun and camouflage paint scheme debarks from a landing craft during the invasion of Cape Gloucester. At first glance, this vehicle appears to be an M3 gun motor carriage, which was based on the M3 half-track. However, a closer inspection reveals that the gun and shield are mounted on an M2 (or possibly M4) half-track. In addition to two .50-caliber machine guns, a .30-caliber machine gun is on a pintle mount to the right of the 75mm gun barrel. (USMC)*

Above: *This USMC M2 has just come ashore in the Marshall Islands on February 1, 1944. It is being used as a communications vehicle and has its fill protective tarpaulin in place. It tows a 1-ton Ben Hur trailer, which would be equipped with a generator and fuel storage to power the radios.* **Below**: *An M2 of an unknown anti-tank company on New Britain Island in 1944. The crew is taking a break under a simple lean-to, while a remaining crewmember maintains a radio watch. The .50-caliber weapon mounts a high capacity "tombstone" 200-round ammunition drum, which would greatly increase the volume of fire generated by this vehicle. (NARA, both)*

Above: *A Japanese anti-tank mine has disabled this M2 Half-track on Bougainville, Solomon Islands on April 1, 1944. This vehicle is camouflaged with olive drab and dark green. Note the 100-round drum-type cartridge on the M1928A1 Thompson sub-machine gun leaning against the bumper of this winch-equipped M2 Half-track. (NARA)*

Below: *P-47 fighter planes of the 318th Fighter Group, returning from a mission over Japan, pass over an M2A1 half-track guarding their home airfield in Okinawa. The vehicle was deployed in an anti-paratroop role to defend against Japanese airborne troops should they attempt to retake the base. (NARA)*

Above: *Another M2A1 conducts forward reconnaissance at Naha airfield on Okinawa, 12 June 1945. In the background is an abandoned Nakajima Ki-84 Hayate fighter.* **Below:** *This M2A1 half-track, registration number 4078817, at the Bengal Air Depot, India, was put to use as a rocket-launcher vehicle and would have been used in an antitank or anti-vehicular role. It includes two sets of three launching tubes for 4.5-inch infantry rockets with high-explosive warheads, on what appears to be a movable mount. The words "NO RADIO" are stenciled on the side of the body, to the rear of the star. (NARA, both)*

Chapter 6
The M3 Personnel Carrier

Diamond T and Autocar Make a Big Push

While the M2 was designed to serve as an artillery prime mover and reconnaissance vehicle, a sister vehicle, the M3, was conceived as a personnel carrier for the armored infantry. While the forward armor and power train of the M3 was identical to that of the M2, the rear body was ten inches longer and had a very different layout. The M3 eliminated the ammo stowage compartments, along with their associated side doors, but added a door in the rear of the vehicle to permit the infantry easy entrance and egress from the vehicle. Rather than a skate rail encircling the fighting compartment, which would have partially blocked the rear door, an M25 pedestal mount with D54075 cradle was utilized to support the standard M1919A4 .30 caliber machine gun. The self-sealing fuel tanks were relocated from the rear of the vehicle forward, to the positions occupied by the ammo chests of the M2. Whereas the M2 had seating for ten, the M3 could accommodate 13.

The Personnel Carriers underwent the same evolution in mine racks, headlights, suspension, fuel filters, etc., as did the M2, the details of which are discussed in the preceding chapter. Likewise, the machine gun mounting on the M3 evolved to the ring mount above the assistant driver's seat in October 1943, with the improved vehicle being designated M3A1.

Diamond T delivered its last M3 in November 1943, when production shifted to the M3A1, the last of which was produced in February 1944. White delivered its last M3/M3A1 in November 1943. Autocar delivered its last M3 in October 1943, and its last M3A1 the next month, but work converting M3 Gun Motor Carriages to M3A1 Personnel Carriers went on into April 1945. While most Lend-Lease transfers were of International Harvester-produced half-tracks, records indicate that 1,475 M3-series Personnel Carriers were transferred to Free French Forces, and a more modest quantities of 40 to Canada; three to Brazil; two each to Great Britain and the Soviet Union. These figures do not include Theater transfers, nor those transferred following the end of WWII.

In August 1942 the Army Ground Forces recommended to the Cavalry Board that the M2 and M3 be consolidated into one vehicle. In response, the President of the Cavalry Board, Colonel H. L. Flynn, offered a comparison of the types. The advantages of the M3 were cited as increased personnel capacity, larger body permitting mounting of assault guns, the convenience of the rear door and a more efficient anti-aircraft gun installation. The M2, on the other hand, was noted as having greater firepower (two .30 caliber and one .50 caliber machine guns vs. the single .30 caliber for the M3 at the time of Flynn's writing, although this would be reduced by one .30 caliber machine gun by the end of the month), and the presence of radio equipment, lacking in the M3. Thus, Flynn concluded, the M3 could be substituted for

M3 half-track personnel carriers are in production at the Diamond T plant. Two of the vehicles to the right have been marked with U.S. Army registration numbers W-401997 and W-402003. In the background are hundreds of parts crates and vehicle components. (NARA)

AMERICAN ROOF TRUSS CO.

the M2 by incorporation of radio gear and augmenting the armament.

Following up on this, on 7 January 1943 authority was granted to divert six M3 carriers to provide for the development of one universal body type to replace the M2 and M3. Interestingly, although not involved in the production of either type, a contract was negotiated with International Harvester to modify the vehicles according to general arrangements outlined by the Armored Force Board in Project 302-1, thereby making these six vehicles of the type.

OCM 20070 22 March 1943 designated the consolidated M2 and M3 the T29. The first T29 pilot was completed on 30 April 1943 and subsequently approved by the Armored Force Board on 29 May. Four more pilots followed, two going to Fort Knox and one each to Autocar, Diamond T and White. The latter three were to be used as manufacturing samples, while the two vehicles at Fort Knox were to be used for study training purposes.

Accordingly, on 20 July 1943 OCM 21501 recommended that the M3A2 be classified as standard and on 30 April OCM 20438 recommended the M5A2, previously known as the T31, be classified as substitute standard. The same Ordnance Committee action reclassified the M3 and M3A1 to Limited Standard type. This recommendation was approved on 2 October 1943, and on 22 December 1943 the initial engineering release for the M3A2 was made. By that time, however, half-track production had ceased. The Ordnance Department noted that at such time as half-track production resume, it would be in the form of the M3A2. That time never came.

The Korean War brought about an increased demand for half-tracks equipped with quad .50-caliber machine gun mounts. To fill this need, Bowen and McLaughlin were contracted to convert 1,662 M3 personnel carriers to multiple gun motor carriages. The resultant vehicle was classified Substitute Standard by OCM 34189 on 24 April 1952, which also designated the vehicle the M16A1. Through this program and attrition, by 10 February 1955, U.S. Army stocks of the M3 and M3A1 had dropped to 1,518 examples worldwide.

Above: *Similar in appearance to the M2 half-track car, the M3 half-track was designed as a personnel carrier. The M3 and its derivatives were put to varied uses during and after World War II, most notably as a conveyance for the armored infantry units that supported tanks. Shown here is an early M3, registration number W-4015741, equipped with a front roller. (ATHS)*

Below: *The sprung bearing of the roller is pictured in this view of the same early M3. The headlights are the early, non-removable style with large, inverted U-shaped brush guards. Small blackout marker lamps are to the lower outboard side of each headlight. Directional-tread tires are in use. (ATHS)*

U.S.A.W-40

W-4015741

Above: *From the front, the M3 half-track is virtually indistinguishable from the M2. A close inspection of this vehicle reveals that the M3 lacks the skate rail of the M2, depending instead on a pedestal mount toward the front of the crew compartment for supporting a .30-caliber or .50-caliber machine gun. (NARA)*

Below: *When viewed from the side, several differences between the M3 and the M2 become apparent. The M3 lacks the ammunition locker doors to the rear of the cab doors. The body of the M3 is 10 inches longer than that of the M2, and the rear is flush with the triangular gussets. (ATHS)*

Left: *With the rear door of the same early M3 open, a glimpse is provided of the crew compartment and the five inward-facing seats along the left side. The latching mechanism for the rear door and its link rods are externally mounted. Below the door is a towing pintle. (ATHS)*

Below: *The canvas tarpaulin over the cab and crew compartment of the M3 half-track is secured with webbing straps cinched to footman loops around the upper part of the body. A large flap over the door provides extra clearance for troops entering or exiting the vehicle, and smaller flaps are on the sides of the tarpaulin. (ATHS)*

Above: *The upper plates of the bodywork around the crew compartment of the M3 lack the numerous slotted, oval-headed screws used on the M2 half-track to fasten the skate rail to the body. The fenders of the M3, identical to those of the M2, are rather complex in shape; these were simplified considerably on the International Harvester M5 and M9 half-tracks. (ATHS)*

Below: *The same M3 half-track, photographed after the installation of the tarpaulin cover. When the cover is on, rather than closing the armored upper panels of the doors, canvas panels, each with a clear plastic window, were used. Note how the rod on the upper door panel locked the panel to the latch bracket on the side of the door. (ATHS)*

Above: *This M3 half-track exhibits several late-model features, including an antitank mine rack on the side of the body, a 5-gallon liquid container rack on the side of the cowl, and removable headlights with revised brush guards. Barely visible next to the idler is the late-type idler spring. (ATHS)*

Left: *An overhead view of an M3 taken at Aberdeen Proving Ground on 27 June 1941, illustrates the layout of the crew compartment, including the pedestal mount for the machine gun. The fuel tanks are to the rear of the front seats, as opposed to the rear of the crew compartment on the M2 half-track. Three seats are arranged side by side in the cab. Stowage areas and rifle racks are behind the three rear seats on each side. (NARA)*

An overhead view of an M3 half-track shows the arrangement of the seats: three in the cab and ten in the rear. To the rear of the cab is a pedestal for a machine gun. To the rear of each fuel tank is a rack for stowing multiple rifles, as well as open stowage bins. (Ordnance Museum)

Above: *This late-model M3 half-track produced by Autocar, registration number 4065403, was photographed in August 1943. When mine racks were installed on M3 half-tracks, they extended from the rear of the side doors to the rear of the body. The mine racks on M2 half-tracks extend only to the rear of each ammunition locker door. (TACOM LCMC History Office)*
Below: *The M3A1 is fundamentally an M3 half-track with an M49 ring mount for a machine gun installed above the right front passenger's seat. This example, registration number 4063674, was produced by Diamond T and photographed at the Ordnance Operation, Engineering Standards Vehicle Laboratory in Detroit, Michigan, on 16 February 1944. Note that the lower left corner of the right windshield has been shattered. (TACOM LCMC History Office)*

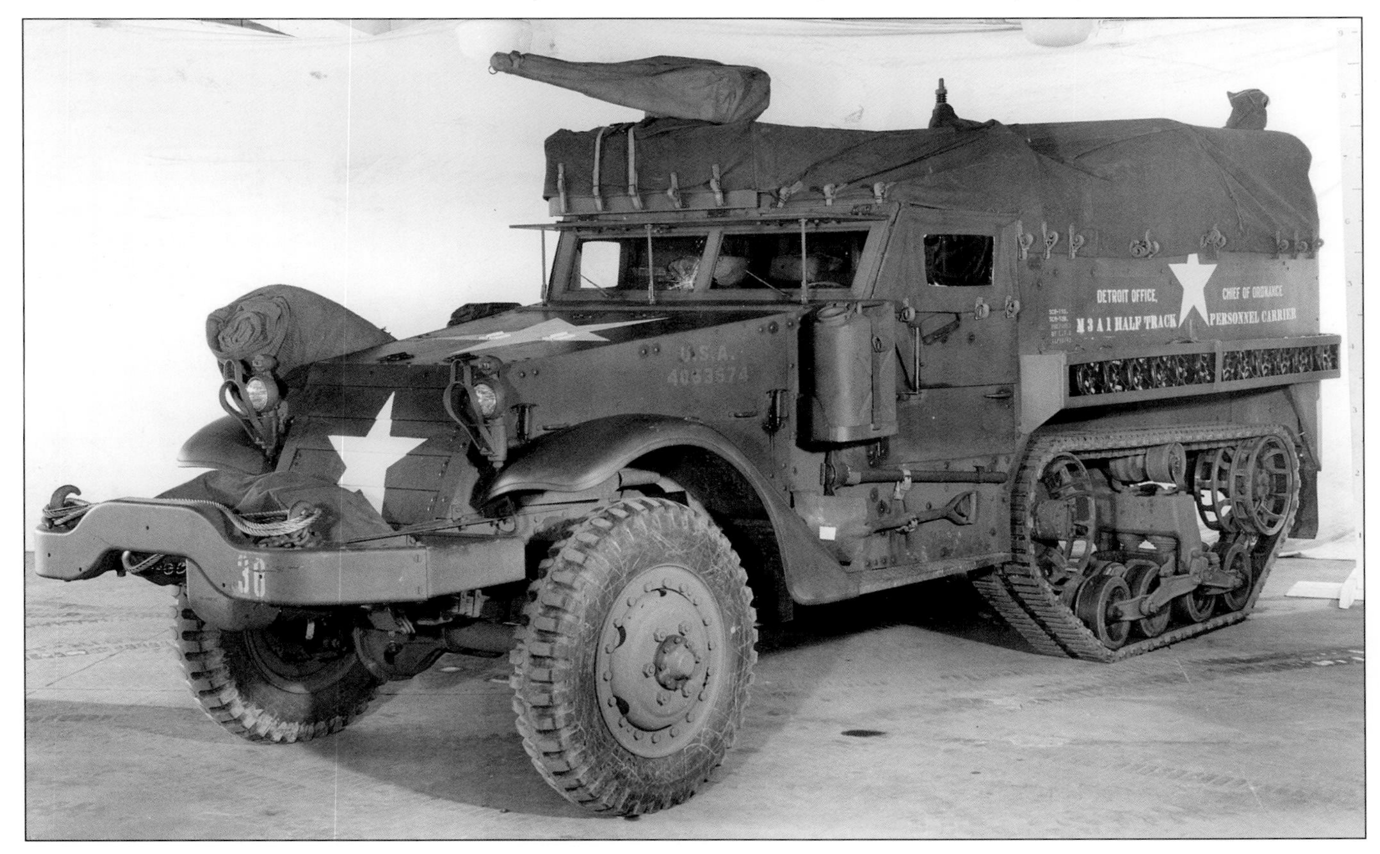

Above: *The same M3A1 photographed with the tarpaulin cover removed and stowed on the left fender. A Tulsa Model 18G 10,000-pound winch is mounted on the front of the chassis frame. The late, detachable headlights with blackout marker lamps on top are present. The headlight brush guards are much simpler and more form fitting than the early brush guards. (TACOM LCMC History Office)*

Below: *As on the M3, the rear bumpers of the M3A1 are simply steel channels fastened with hex bolts to the bottom of the rear of the body. Bows are stowed in the handholds, and an M2 tripod for a .30-caliber machine gun is inside the canvas bag strapped to the rear door. The rear ends of the mine racks protrude from the sides of the vehicle. (TACOM LCMC History Office)*

Above: *The right side of this M3A1, photographed at the Engineering Standards Vehicle Laboratory in February 1944, is shown. In addition to the Browning .50-caliber M2HB machine gun on the M49 ring mount, a Browning M1919A4 .30-caliber machine gun is on the right side of the crew compartment. Each rack could hold 12 antitank mines. (TACOM LCMC History Office)*

Below: *The relative locations of the two radio antennas are illustrated in this photo. The rolled, face-hardened armor of the M3A1, like that of the M3, is ¼-inch thick, with the exception of the windshield cover and the sliding visor plates on the upper panels of the doors, which are ½-inch thick. (TACOM LCMC History Office)*

Above: *The same vehicle is shown from the same angle with the tarpaulin installed. Canvas flaps are pulled open to allow the radio antenna mounts to protrude. The M2HB .50-caliber machine gun has its own fitted cover. The ends of the tarpaulin straps are neatly rolled up.* **Below:** *The M3A1 under evaluation at the Engineering Standards Vehicle Laboratory was also photographed from overhead. A notable difference between the M3A1 and the M3 is the addition of three fixed sockets for pintle-mounted .30-caliber machine guns in the crew compartment: one on each side and one on the rear. This vehicle has a metal cover over the stowage bin behind the left rear seats. (TACOM LCMC History Office, both)*

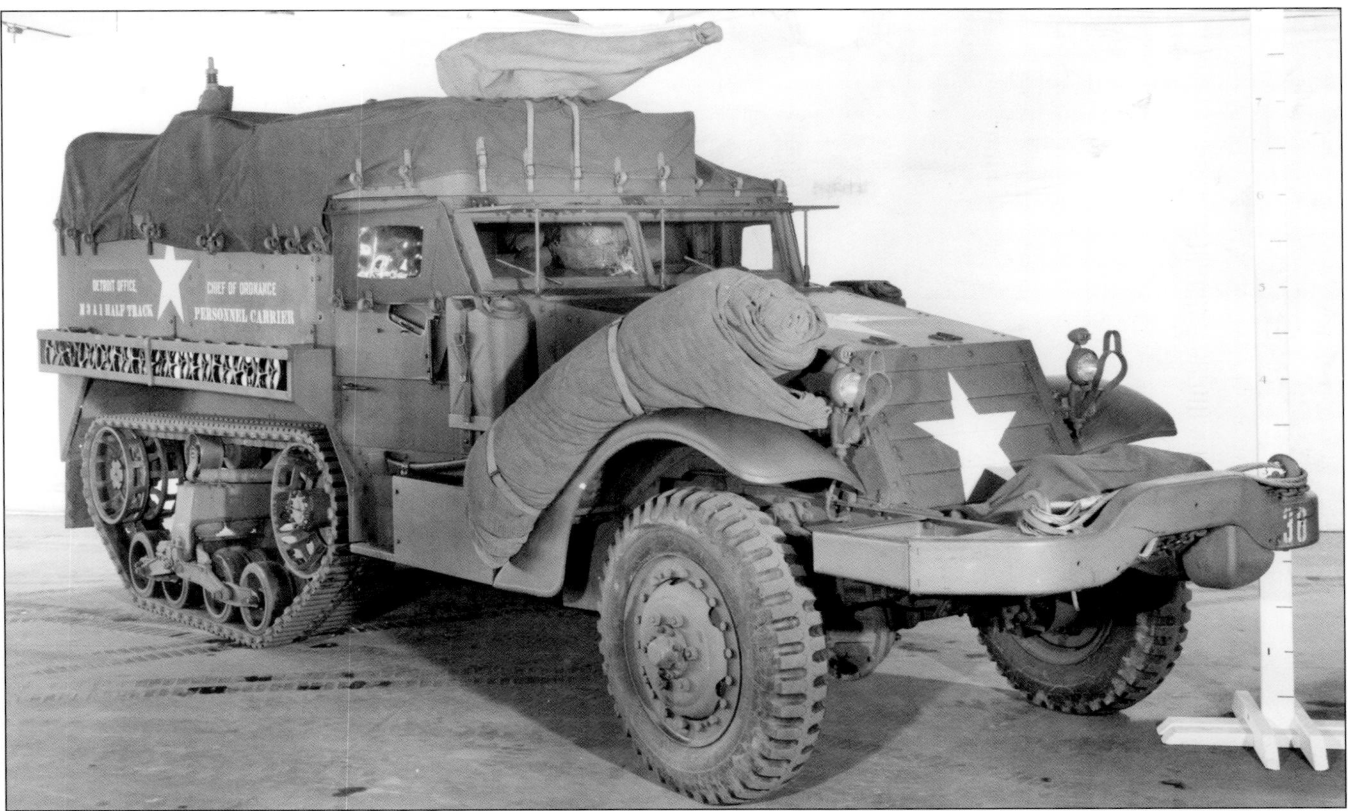

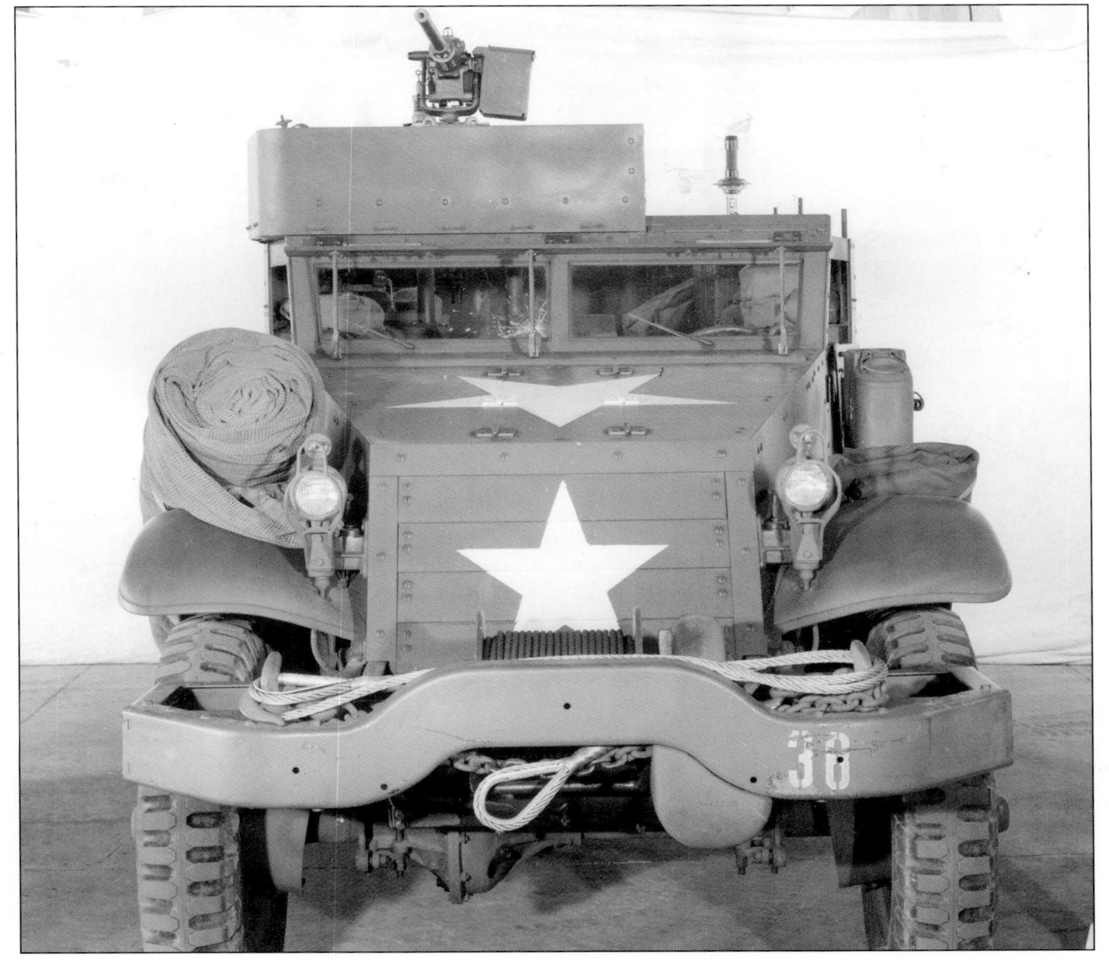

Above: *The M3A1 is shown from the front right with tarpaulin on and radiator shutters closed. The 5-gallon liquid container is secured to its rack with a webbing strap. Note the small loops on top of the headlight brush guards, which coincide with the small marker lamps on top of the headlights. (TACOM LCMC History Office)*

Below: *Three steel rods support the armored cover of the windshield when open. The rods could be manually detached and the cover lowered by hand when the vehicle was "buttoned up." Below the bumper is the front of the gear case of the Tulsa winch. Also visible from this angle is the front differential and axle. (TACOM LCMC History Office)*

Above: *The engine compartment of the M3A1 half-track was virtually identical to that of the M3 and, for that matter, the M2 and M2A1. The centerpiece is the White 160AX engine. The placard on the surge tank to the right reads, "Never remove this cap. Use radiator cap for filling. WARNING: Do not remove radiator cap when engine is hot." (TACOM LCMC History Office)*

Below: *Elements in the right side of the engine compartment of an M3A1 include, left to right, the oil-bath air cleaner, air hose, carburetor, crankcase oil filler, spark plugs, and oil filter (a FRAM filter body No. 5267). On top of the oil filter is a label providing servicing instructions. (TACOM LCMC History Office)*

Left: *Folded blankets were inserted into all 13 of the zippered canvas seat covers on the M3A1 for cushioning as well as economy of stowage. The center front seat cushion has been removed. On this vehicle, an intercom box has been installed to the right of the glove box for the use of the squad leader. (TACOM LCMC History Office)*

Below: *Looking through the rear door of a Diamond T M3A1, inward-facing seats for ten men are to the rear of the three front seats. To the far right is a radio set. In the background is the tubular support for the M49 ring mount. The floor plates are removable, providing access to stowage spaces for vehicular tools and equipment. (TACOM LCMC History Office)*

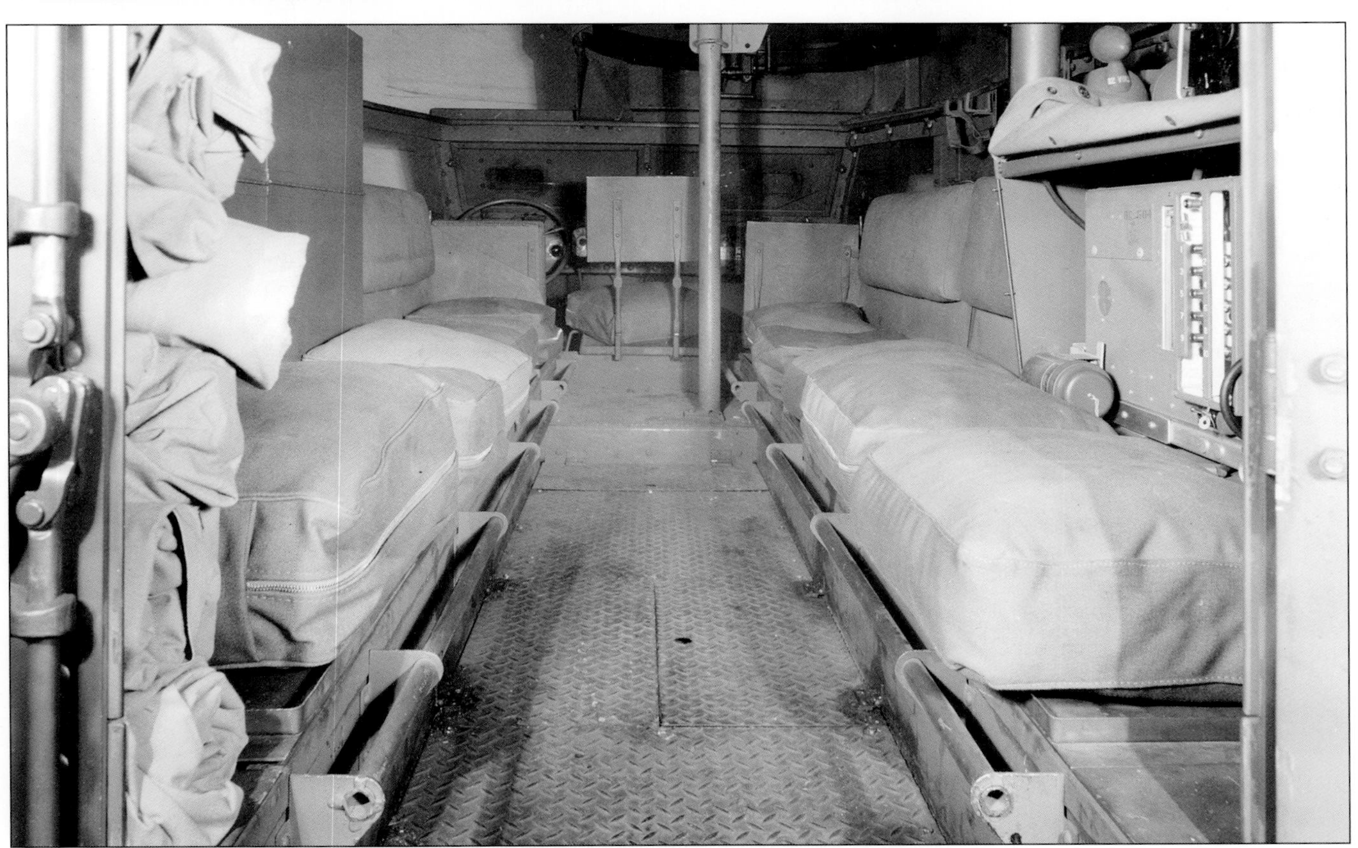

Above: *Three officers and two civilians examine the engine compartment of a newly built M3 half-track with the tarpaulin cover in place. Eight webbing straps across the front of the tarpaulin secure that portion of the cover to footman loops on top of the windshield frame. Note the remarkably clean tires. (ATHS)*

Below: *This M3 half-track with registration number W-409485 painted in blue drab on the engine compartment cover is bristling with stowed M1 Garand rifles with bayonets affixed. All three bows are installed: the rear bow is tilted toward the rear, while the other two support bows are vertical. (NARA)*

Above: *In the same M3, a full complement of thirteen G.I.s demonstrate the seating arrangement in that vehicle, with ten men in the rear facing inward and three, including the driver, to the front. Conditions were cramped and would be even more so when the vehicle was deployed with full combat stores. Varying numbers of crewmen were assigned to half-tracks depending on the mission and the vehicle's configuration. (NARA)* **Below:** *The crew of an M3 half-track stand with their vehicle (with the nickname "CABOOSE" painted below the driver's door) and its on-vehicle equipment, including mines, pioneer tools, cables, mechanical tools, and other gear. (Patton Museum)*

Above: *This photo, taken by the Tank Destroyer Board at Fort Hood in February of 1943, illustrates the proper stowage for the M3 when used as a prime mover for the 3-inch M5 anti-tank gun. The complement was a full 10 men, plus gear, individual equipment, weapons and 32 rounds of 3-inch ammunition. (NARA)*

Below: *An Aberdeen Proving Ground photograph dated 25 August 1942 shows an experimental installation of ice scrapers abutting the sprocket and the idler of an M3 half-track. These were steel plates attached to pipes jutting from the chassis frame. (NARA)*

Above: *The same M3 half-track with the ice scrapers is shown after the vehicle had been driven through a muddy test course at Aberdeen. The scrapers cleared mud from the idler and sprocket but undesirably dumped mud onto the tops of the bogie wheels. (NARA)*

Below: *An M3 half-track, left, and an M2 half-track drive onto a partially constructed pontoon bridge, perhaps to test the loading of the bridge. The photo probably was taken during 4th Armored Division maneuvers in the fall of 1941. Indistinct round insignia are on the half-tracks. (NARA)*

The original caption states that this is T5 Keith L. Shumpert (6-feet, 6-inches, 260 pounds) helping Private Jamey Driver (5-feet, ½-inch, 139 pounds) into an M3 half-track. The photo was taken during the Louisiana maneuvers of August 1942. Note the Springfield M1903 rifles. (NARA)

Above: *A civilian train conductor prepares to turn over responsibility for his cargo on 1 November 1941. The 1st Armored Division had just arrived in Rock Hill, South Carolina to participate in the Carolina maneuvers. The two-star placard on this M3 indicates it is the mount of the division commander, Major General Bruce Magruder.* **Below:** *A column of armored-infantry half-tracks led by an M3 proceeds along a country road during First Army maneuvers in South Carolina on 6 November 1941. The roller on the M3 has a tight-fitting cover on its left side, giving the roller a two-tone appearance. (NARA, both)*

Above: *An M3 half-track passes an M4 medium tank during maneuvers at the armored force school at Fort Knox, Kentucky, in July 1942. Armored infantry units, operating from M3s and their derivatives, would act in support of U.S. tank forces in Europe, forging a formidable combination.* **Below:** *At Fort Knox in June 1942, an M3 passes along a hot, dusty trail. The crewmen are still wearing M1917 helmets, surplus from World War I. Just visible is the muzzle end of a water-cooled M1917A1 .30-caliber machine gun with its water hose. (LOC, both)*

Above: *An M3 half-track passes a parked M3 medium tank during training exercises at Fort Knox in June 1942. One of the bows for the tarpaulin cover is installed. The roller appears to have taken quite a beating during recent operations.* **Below:** *A column of half-tracks, including an M3 in the lead, conduct training exercises at Fort Benning in April 1942. Protruding above the sides of the crew compartment are extension rods, to which the bows are affixed when the tarpaulin cover is installed. These rods were normally stowed behind the left rear seats. (LOC, both)*

In a photo taken during the Tennessee maneuvers of late 1942, an M3 half-track with markings for Headquarters, 37th Armor Regiment, 4th Armored Division, is among vehicles pausing along a rural road. Knapsacks hang from the single bow that is installed. (NARA)

Above: *An M3 half-track of the 12th Armored Division slogs through soupy mud in a forest during maneuvers. A temporary tactical symbol is on the body. Interestingly, slices of a tree trunk have been installed in the mine racks to replicate mines. (NARA)*

Below: *During the Tennessee maneuvers of late 1943, an M3 half-track of the 17th Armored Infantry Battalion, 12th Armored Division, was photographed during a night march. The tactical symbol on the body of the vehicle is a large X, with a small 1 or I over an A.*

Above: *A half-track is loaded on a railroad flatcar for shipment in late 1942. Note the cover over the radiator-armor flaps. (NARA)* **Below:** *During exercises at the Desert Training Center in Southern California in 1942, an M3 is put through its paces negotiating a rugged arroyo. Racks for carrying 5-gallon liquid containers have been added to the rear of the vehicle on each side of the door. The name "Warrior" is stenciled at the bottom of the driver's door. (NARA, both)*

Above: *Half-tracks, including an M3 in the foreground, are lined up at the Desert Training Center in California. The M3 is an early-production example, with early-type headlights and brush guards, and without the idler springs of the later-model M3s.*
Below: *M3 half-track 4027980 is equipped with radio equipment. A whip antenna is installed on the pedestal at the front of the crew compartment, two more antennas are toward the rear, and what appears to be radio equipment protrudes above the top of the crew compartment. The scene probably was the Desert Training Center in the Mojave Desert. (NARA, both)*

Above: *While the ever-flamboyant Gen. George S. Patton was commanding the I Armored Corps at the Desert Training Center in 1942, he had this M3 half-track, U.S.A. number 402408, customized with a hardtop with a .30-caliber machine gun embrasure, side skirts, and radio equipment. "NITE OWL" was painted on the bottom of the driver's door. (NARA)*

Left: *Patton's personal M3 half-track at the Desert Training Center is viewed from the rear. On the left bumper was a tactical sign for an armored unit and "HQ-1," and on the right bumper was "1-56." (NARA)*

Above: *A weather-beaten M3 half-track with a cargo of G.I.s passes through Mazagan, on the coast of French Morocco during Operation Torch in late 1942. For the Torch landings, the vehicle was fitted with a fording kit, including the exhaust extension. For identification purposes, a U.S. flag has been applied to the side of the half-track, and an indistinguishable name is stenciled on the bottom of the door. (Patton Museum)* **Below:** *G.I.s load 105mm ammunition in packing tubes into an M3 half-track. The vehicle is being used to shuttle ammunition to the 27th Armored Field Artillery in February of 1943. The 27th AFA was equipped with M3-based T19 105mm HMCs. (NARA)*

Above: *During the Battle of Kasserine Pass in Tunisia on 24 February 1943, elements of the 6th Armored Infantry, 1st Armored Division, take cover by a mud hut and watch for enemy aircraft. Among the vehicles present is an M3 with what appears to be mud camouflage applied. (NARA)* **Below:** *General Erwin Rommel and Lieutenant Colonel Fritz Bayerlein pass a captured M3 half-track. The Germans "inherited" nearly 100 half-tracks of various types during the Kasserine-Faid Pass battles of February 1943. Always short of motor transport, these vehicles quickly supplemented their dwindling stock. (BA 1990-071-31)*

Above: *This M3 half-track was photographed in Oran, Algeria, on 6 April 1943. Although U.S. half-tracks had their introduction to combat in the Philippines and at Guadalcanal in 1942, their first large-scale operational use was in North Africa. (NARA)* **Below:** *The Germans captured this M3 half-track during the Tunisian Campaign and were using it to lead a counterattack at El-Guettar when U.S. forces knocked it out. The nickname "DOLORES" is faintly visible on the lower part of the driver's door. (ATHS archives)*

Above: *Members of C Company, 82nd Reconnaissance Battalion, 2nd Armored Division, perform a security patrol in an M3 half-track on a street in Ribera, Sicily, on 25 July 1943. A name is partially visible on the bottom of the driver's door, with the letters "COGHRA" being visible, preceded and followed by other letters. (NARA)* **Below**: *A British M3 half-track personnel carrier drives onto a landing-beach carpet at Reggio, Italy, on 3 September 1943. Numerous unit and formation markings and placards are on the vehicle, as well as a red-white-red recognition flash on the side. (IWM)*

Above: *A very unusual stowage bin fabricated from a steel grille has been mounted on the side of this M3 half-track to hold the crew's equipment. Other unique touches include the fire extinguisher to the rear of the driver's door and the machine gun on the rear door. (Mike Peters)*

Below: *A Diamond T M3 half-track, registration number 4045987, undergoes repairs on its rear suspension in an ordnance base shop at Leghorn (Livorno), Italy, sometime during 1944. Screw jacks are supporting the rear end, and the left suspension and continuous rubber-band track have been completely removed. (Patton Museum)*

Above: *A farmer leads his horse past a field full of U.S. Army vehicles somewhere in England in March 1944. Most of the vehicles are half-tracks, with several M3s being discernable. All are fitted with their canvas covers to keep out the elements. (NARA)*

Below: *Half-track 4062301 exits the Utah beachhead with a variety of vehicle types following it. "LST 1009" and "A23" are visible on the bumper. A crate is stowed below the machine-gun pulpit, and large rolls of barbed wire are on the bow and the side of the crew compartment. (NARA)*

Above: *A week after D-Day, on 13 June 1944 an M3A1 half-track towing an antitank gun rolls through a village in Normandy. Four extra liquid containers are stowed in a rack above the mine rack. (NARA)*

Below: *Residents of Sartilly, France, wave at U.S. troops in late July 1944. The half-track is M3A1 U.S.A. number 40102907-S, the "S" meaning that the vehicle had passed a radio-interference suppression test. (NARA)*

Above: -*A 3rd Armored Division convoy led by five half-tracks crosses a pontoon bridge over a river in France on 26 August 1944. Jutting over the windshield of the first vehicle, an M3A1, is a rack for extra ammunition. Items stashed on the vehicle include a toolbox and vise on the left side of the winch. (NARA)*

Below: *Half-tracks assigned to an armored engineer unit cross a pontoon bridge at Pointe-la-Rouge, Normandy, on 31 July 1944. Ammunition boxes are stowed in the mine rack, and a roll of camouflage netting is on each fender. (NARA)*

Above:: *A U.S. half-track rolls over the Treadway Bridge that U.S. Army engineers erected over the river at Gavray, France, hours after American forces captured the town on 1 August 1944. The Germans destroyed the town bridge before they left in an attempt to halt the American advance. (Image Bank WW2)*

Below: *A column of U.S. Army half-tracks of the 4th Armored Division advances through a ruined Avranches on 1 August 1944. The lead vehicle is marked "HOLY-DEVILS" on the driver's door. Flanking the identification star on the side of the crew compartment are two markings, a triangle and a partially obscured number 8. (NARA)*

Above: *In late August 1944, an M3 half-track rolls through newly liberated Paris. A section of Pressed Steel Plating (PSP), used for constructing temporary landing strips, is being used as a rack for liquid containers and other gear.* **Below:** *American armored forces of the 4th Armored Division roll down the road near the hamlet of Piney, France on 27 August 1944. Piney sits on the road northeast of Troyes and about 200 kilometers from the town of Lunéville, which the 4th AD would take less than three weeks later. Evidence of American airpower is strewn along the roadside. (NARA, both)*

Above: *Smiling townspeople watch as two U.S. M3A1 half-tracks advance through Catigny, France on 31 August 1944. A military censor blotted-out the unit marking on the right side of the bumper, but the alphanumeric code D-9 remains visible on the left side of the bumper and in large figures on the side of the crew compartment. The original caption tells us that the vehicles are members of Company D, 41st Armored Infantry, 2nd Armored Division.* **Below:** *Engineers construct a plank bridge near the west end of Cannes in southern France on 26 August 1944. The Germans had destroyed the existing bridge, but it is not apparent from the original caption how the M3 half-track in the center of the photo ended up on its side. (NARA, both)*

Below: *Wearing a communications headset, Sergeant Fred Scott, left, delivers a verbal message to a German machine-gun squad to surrender using a loudspeaker-equipped M3 half-track at Eilendorf, Germany, on 23 September 1944. (NARA)*

Below: *Extensive stenciling is present on the side of this M3A1 half-track in a ruined town in the European Theater. Additional faint stenciling is barely visible to the rear of the identification star. (NARA)*

Below: *An M3A1 half-track personnel carrier of the 7th Armored Division rolls through Saint-Vith, Belgium, 13 September 1944. A good view is available of the tarpaulin covering the machine-gun pulpit and the body. (NARA)*

Below: *Members of a wire section of the 6th Armored Division lay telephone wire from an M3 half-track in France in October 1944. All sorts of vehicles were used for laying wire, but a half-track offered armored protection and cross-country mobility. (NARA)*

Below: *These two knocked out U.S. Army M4 Sherman tanks and an M3A1 half-track were destroyed in a battle near Nancy, France, in October 1944. A large hole was ripped in the side of the half-track, and the tires and tracks were burned to ashes. (NARA)*

Below: *A half-track heavily laden with baggage drives through a gap between "dragon's teeth" antitank obstacles somewhere in northwestern Europe on 19 November 1944. Bags and rations boxes are secured to the top of the raised windshield armor. (NARA)*

Above: *A burned-out M3A1 sits along a street in Mulhouse, France, 22 November 1944. It caught fire during fighting for the Gestapo headquarters in that town when a German rifle bullet pierced a gasoline can on the half-track, igniting the contents. (NARA)*

Below: *Hot chow by their half-track is the order of the day for these members of an ammunition-carrier section near Brulange, France, on 15 November 1944. They had just finished transporting a load of ammunition up to the front lines. (NARA)*

Below: *An M3A1 half-track with a boom rigged to the front bumper is being used to recover a Jeep. The winch at the front of the half-track operated the hoist cable. Two stabilizing jacks are employed underneath the bumper.* **Below:** *A U.S. Army M3A1 half-track drives past a destroyed Sherman tank in Barre, France on 29 November 1944. Bedrolls are piled up in front of the right side of the armored windshield. An inscription, "Baby Bastard No. 1," is painted at the top of the radiator armor. (NARA, both)*

Above: *An M3A1 half-track of the 4th Armored Division approaches the body of an American who was killed in action near Chaumont, Luxembourg, during the Battle of the Bulge in late December 1944.* **Below:** *In another photo taken near Chaumont, Luxembourg, in late December 1944, an M3A1 of the 4th Armored Division passes the same destroyed jeep. On the rear of the half-track is a stowage rack made of welded pipes. (NARA, both)*

Above: *Two elderly refugees struggle past a knocked-out M3 or M3A1 half-track near Bastogne on 27 December 1944. A stowage rack fabricated from angle irons has been blown off of the rear end of the half-track.* **Below:** *Members of the 11th Armored Division prepare to attack German forces on the outskirts of Bastogne, Belgium, on 31 December 1944. In the left foreground is an M3A1 half-track with knapsacks neatly stowed above the mine rack. (NARA, both)*

Above: *An M3A1 half-track named "Agnes IV" lies on a road, the front right wheel blown off, most likely from a mine, judging by the torn-up condition of the pavement. Markings for the 9th Armored Division are on the bumper.* **Below:** *An M3 half-track full of troops of the 4th Armored Division passes a group of German prisoners of war on the outskirts of Bastogne, Belgium, on 27 December 1944. Rickety-looking racks on the rear of the vehicle hold rations boxes and packs. (NARA, both)*

Above: *Under camouflage netting, G.I.s perform tire changes next to an M3-type half-track with "DER YANK" with a picture of a bulldog painted on the cab door. Above the mine rack are a pioneer tool rack and another rack for stowing equipment.*
Below: *M3A1 half-tracks of an undisclosed unit plow through the snow near Born Belgium on 21 January 1945. (NARA, both)*

Above: *On 14 January 1945, a half-track of the 8th Armored Division passes a knocked-out German PzKpfw IV near Bastogne, Belgium. The whitewashed half-track was an M3 modified to M3A1 standards by installing an M49 ring mount. The early-style, fender-mounted headlights and brush guards have not been replaced by their late-type counterparts. The roller assembly is missing from the bumper.* **Below:** *U.S. Army half-tracks are parked in a snowy field during the final winter of World War II. In the left foreground is an M3A1. All of the vehicles have been camouflaged with white paint or whitewash to make them blend in with the snowy terrain. (NARA, both)*

Above: *Staff Sergeant George Brackemyre, left, of Seymour, Indiana fires the .50-caliber machine gun on his half-track at a German sniper position in Saarburg, Germany, in February 1945. A German sign is strapped to the tarpaulin.* **Below:** *Nuns in a building in Rouffach, France, look down on U.S. troops taking a break after helping seal the Colmar Pocket in February 1945. To the right is a half-track of the 119th Engineers, 12th Armored Division, equipped with two large stowage bins on the rear. On the side of the crew compartment is a tactical sign composed of a vertical bar with a circle on each side. (NARA, both)*

Above: *A G.I. mans the .50-caliber machine gun of a half-track as it passes through Remscheid, Germany on 28 February 1945. Next to the soldier, protruding through the tarpaulin, is a radio antenna and its base.* **Below:** *In a war-torn town in the European Theater, an M3A1 half-track is parked adjacent to a Jeep and two M4A3 Sherman tanks: the one to the left is an M4A3E2 "Jumbo" assault tank. An improvised stowage rack is on the rear of the half-track's body. (NARA, both)*

Above: *An M3 half-track towing an artillery piece passes through a gate in Colmar, Germany on 2 February 1945. This vehicle and its crew are part of the French 2nd Armored Division. Snow chains are installed on the tracks but not on the tires. (NARA)*

Below: *The gunner in the pulpit of a half-track fires the machine gun at a target in the background while the G.I. to the left observes the effects of the fire with binoculars. Interestingly, the mines stored on the rack on the side of the crew compartment are individually wrapped. (NARA)*

Above: *In a town in the ETO congested with U.S. Army vehicles, several half-tracks are visible. Many different designs of stowage racks were installed on half-tracks in the field, and one such type, stuffed full of bedrolls, is visible close-up on the M3A1 to the far left. (NARA)*

Below: *Captain James L. Anderson, communications officer of the 85th Reconnaissance Squadron, 5th Armored Division, inspects a half-track with a specially built shelter used as the squadron's mobile communications post at Hoensbroek, Holland in February 1945. (NARA)*

Above: *Interesting markings are prominent on this half-track undergoing a .30-caliber ammunition restocking. The insignia of a wolf's head over the yellow, blue, and red armored-division triangle, on a white square, was that of the 15th Tank Battalion of the 6th Armored Division. Vehicles of headquarters company typically had large white numbers ranging from 11 to 19: the significance of the ½ following the 13 on this half-track is unclear.* **Below:** *Members of Company D, 784th Tank Battalion, Task Force Byrne, Ninth U.S. Army, service their vehicles, including the M3A1 half-track in the garage to the right, at Sevelen, Germany, on 5 March 1945. A .30-caliber machine gun is on the pulpit mount. (NARA, both)*

Above: *Half-tracks with trailers hitched of the Third U.S. Army are crossing the Kyll River in Germany on a treadway bridge in early March 1945. The half-track in the foreground is an M3A1. The one on the bridge appears to have International Harvester-type fenders. (NARA)*

Below: *Vehicles of the 5th Armored Division, Ninth Army, proceed through Hardt, Germany, on 21 March 1945. Entering the intersection to the right is M3 half-track U.S.A. number 40102931, with the nickname "ANGEL II" painted on the driver's door. (NARA)*

Above: *Members of a U.S. reconnaissance patrol are depositing a few German POWs at Grevenstroich, Germany, on 3 March 1945. The half-track has markings on the bumper for 24th Cavalry Reconnaissance Squadron, First Army. The half-track and the M8 armored car both have tire chains for added traction. (NARA)*

Below: *In the bombed-out ruins of Steinfeld, Germany, in late March 1945, a U.S. Army sergeant puts up a street sign that reads, "88 AVE / MORTAR / DRIVE!" Behind him is a half-track of the 62nd Antitank Battalion, Combat Command A, 14th Armored Division, with four boxes of K5 rations lashed to the top of the roller. (NARA)*

Above: *Members of the 2nd Battalion, 66th Armored Regiment, 2nd Infantry Division, Ninth Army, grab some rest by their half-track after a night crossing of the Rhine River near Hünxe, Germany, on 28 March 1945. Among the gear stored on the side of the crew compartment are a litter and ammunition boxes.* **Below:** *An M3 half-track assigned to the 10th Armored Division, Third Army, is parked in front of a smoldering building in Kell, Germany, on 18 March 1945. A gunner is at the ready behind a pedestal-mounted .50-caliber machine gun. (NARA, both)*

Above: *Elements of the 9th Armored Division roll into the burning town of Limburg, Germany in March 1945. (NARA)*
Below: *An M3A1 half-track of the 27th Armored Infantry Battalion, Ninth Armored Division, proceeds with windshield armor lowered through Engers, Germany, on 27 March 1945. A name that appears to be "BITCHING PALS" is on the cab door. (Heslop Collection, Brigham Young University)*

A smiling Russian mother and her child walk past an American half-track parked in the yard of an Allied displaced persons camp near Trier, Germany. In the two weeks beginning 5 March 1945 Allied officials dealing with displaced persons handled approximately 23,000 liberated prisoners. (Image Bank WW2)

Above: *A U.S. half-track is at the scene as members of the 125th Engineer Battalion, Combat Command A, 14th Armored Division, Seventh U.S. Army, construct a bridge over the Seltzbach River near Niederroedern, France, on 18 March 1945. A little bit of pinup art is toward the rear of the pulpit machine-gun mount. (NARA)* **Below:** *With the melting of the winter snow, these G.I.s are scrubbing whitewash camouflage off an M3A1 half-track, to better blend in with the springtime environment. The water-based whitewash could be washed off the Olive Drab paint with sufficient elbow grease. (Patton Museum)*

The photographer is about to be photographed by a G.I. sitting in the driver's seat of an M3A1 half-track in a column of vehicles of the 8th Armored Division, Ninth Army, that has paused during a road march toward the Rhine River on 2 March 1945. A clear view is presented of the machine-gun pulpit and the upper frame of the body. (NARA)

Above: *This M3A1 half-track has paint on the machine-gun pulpit and part of the door that is noticeably darker than the base paint of Olive Drab, probably an effort to camouflage the vehicle. A panel with unusual artwork on it is above the front of the mine rack. (NARA)*

Below: *Armored infantry dismount outside of Niederroedern, France to wait for engineers to span the Seltzbach River on 18 March 1945. They are members of the 14th Armored Division, which had just restarted their offensive operations a few days earlier. (NARA)*

Above: *Several local women look on as an M3A1 half-track of the 9th Armored Division proceeds through a German town. The mine rack on the side of the body is being used to stow miscellaneous pieces of equipment. (Heslop Collection, Brigham Young University)*

Below: *German snipers captured by elements of the 3rd Armored Division near Marburg, Germany, are marched past a line of half-tracks on 30 March 1945. The vehicle to the far right, an M3A1, has a stowage rack above the armored windshield. (NARA)*

Above: *Forces of the 104th Division pass through Ebershutz, Germany, during their push into the western bank of the Weser River on 7 April 1945. The M3 half-track has what appears to be an improvised machine-gun ring mount to the center rear of the cab.* **Above:** *The cheerful crew of an M3 half-track of the 46th Armored Infantry, 5th Armored Division, rides past a burning building during the advance across northwestern Europe. "Copenhagen" is written at the bottom of the passenger's door, and an abundance of stowed tarps, bedrolls, crates, ration boxes, and equipment is stowed on board, including on the extra racks on the sides and rear of the vehicle. (NARA, both)*

Above: *A column of vehicles of the 5th Armored Division, Ninth Army, advances toward Bismarck, Germany, on 11 April 1945. In the center background and to the right are M3 half-tracks heavily loaded with baggage on the sides and rears.*
Above: *The crew of an M3A1 half-track observe a building on fire in the background on 17 April 1945. It is a mount of the 61st Armored Infantry, 10th Armored Division. Aft of the .50-caliber machine gun are two .30-caliber machine guns, one of which is air-cooled and the other water-cooled. (NARA, both)*

Above: *During the drive across Europe, as in any military campaign, exhaustion eventually set in among the troops, such as these three G.I.s taking advantage of a lull in action to take a nap on their half-track. A bazooka is lying atop the armored windshield, and a typewriter is behind the man to the right.*

Above: *The body of a G.I. killed in action is being placed onto the hood of an M3A1 half-track of Combat Command A, 14th Armored Division, VI Corps, for evacuation at Steinfeld, Germany, on 23 March 1945. (NARA, both)*

Above: *Toward the end of World War II, stowage racks fitted to the sides and rear of the M3 and M3A1 became commonplace. The ladder-like side racks were used not only for strapping on or stashing gear, but also for securing the straps of the tarpaulin cover. This vehicle has what appears to be a spotty attempt at sprayed-on black camouflage over the base coat of olive drab.* **Below:** *Half-tracks of the 13th Armored Division, Third Army, approach the hamlet of Gschaid, Germany on 2 May 1945. Slightly over 20 miles down this road is the birthplace of Adolf Hitler, Braunau am Inn, Austria. Equipment bins fashioned from welded steel rods are on each side of the rear door and on the side of the crew compartment of the nearest half-track. (NARA, both)*

A driver poses in the cab of an M3 half-track. To the left is a clear view of a removable door curtain with a clear-plastic window. It was fitted with rods that were inserted in the tubular holders on the door. (3rd Cavalry Museum)

Above: *Half-tracks of the 20th Armored Division, Seventh Army, pass through Aichach, Germany, during the advance on Munich on 29 April 1945. Note the variation in the rear stowage boxes, including the full-width one on the second closest vehicle. (NARA)* **Below:** *This M3A1 half-track has a winch-operated A-frame boom, to do heavy lifting. The boom is attached to a projection at the front of the chassis, with two angled support bars and a stay, apparently made of steel strapping, extending to the hood. (3rd Cavalry Museum)*

Above: *An M3A1 half-track with Third Army markings is parked in an alpine area. Mounted on the front face of the machine-gun pulpit is a stowage rack for ammunition boxes. Note the angle iron mounted horizontally above the mine rack. (3rd Cavalry Museum)* **Below:** *As German troops pass along the main street of Schwarzbach, Germany, to surrender, an M3 half-track that had been captured by the Germans is parked to the right. The two angled fixtures on the side of the crew compartment are mounts for short-range, large-caliber rocket launchers, possibly the Wurfrahmen 40. (NARA)*

Above: *President Harry S. Truman, left, and other dignitaries review a line of U.S. Army M5A1 light tanks during the Potsdam Conference in Germany in the summer of 1945. The half-track was nicknamed "BESS" after Truman's wife.* **Below:** *In a war-torn section of Peleliu in the Caroline Islands, a G.I. aims a pedestal-mounted .50-caliber machine gun on an M3 half-track while two soldiers kneel at the ready with M1 rifles. On each side of the rear door was a folding stowage rack. The one on the left was in the folded-up position, while the one on the right holds several boxes. (NARA, both)*

Above: *A G.I. mans the 30-caliber machine gun as his M3 half-track of the 37th Reconnaissance Troup negotiates a muddy, narrow trail on Bougainville in March 1944. The sliding covers on the door and windshield armor panels are closed except for the one in front of the driver, which is slightly open. (NARA)*

Below: *Several M3 half-tracks of the 640th Tank Destroyer Battalion are parked at a base on New Britain on 24 May 1944. The nearest vehicle is U.S. Army registration number 4055112. Small, illegible markings are stenciled on the right side of the bumper. (NARA)*

Above: *Chinese troops of the 1st Chinese Regiment, 5332nd Brigade, 1st Tank Group scramble out of their M3 half-track during training on 25 January, 1945. The location is Kabani, Burma. (NARA)* **Below:** *U.S. members of the 1st Chinese Provisional Tank Battalion mill around their M3A1 half-track during operations in Burma. The American portion of the unit was originally derived from the 527th Ordnance Company (Heavy Maintenance) (Tank). (NARA, both)*

Above: *The M3A2 tested at the Ordnance Operation, Engineering Standards Vehicle Laboratory, Detroit, in September 1944 is shown with the tarpaulin and cab top removed, showing the pulpit with a .50-caliber machine gun on an M49 ring mount. (Patton Museum)*

Below: *An M3A2 is viewed from the left rear, showing the equipment racks loaded with blankets and bedrolls. An M1919-type .30-caliber machine gun is on the rear pintle socket: sockets also were on each side of the body. (TACOM LCMC History Office)*

Above: *The M3A2 half-track cars were the pilots for a proposed replacement for the M2 and M3 half-tracks. Originally designated the T29, the M3A2 was based on the M3A1. M3A2 Ordnance number 4027138 is shown in a 15 September 1944 photo. (TACOM LCMC History Office)*

Below: *The M3A2 had stowage racks, which resemble ladders, above the mine racks, the top rails of which doubled as tie-down points for the tarpaulin. This vehicle was manufactured by Autocar, and International Harvester performed the work to modify it to a T29/M3A2. (TACOM LCMC History Office)*

Above: *An M3A2 is viewed from above, providing an excellent view of the top of the machine-gun pulpit and ring mount as well as of the two rear stowage racks. Also in view are the side and rear pintle sockets and the seats. (Patton Museum)*

Below: *An SCR-506 radio set is installed in the left rear of this T29 (later redesignated M3A2) half-track. Two antenna-base assemblies are present. To the rear of the radio are rifles in canvas covers. A large number of ammunition boxes are stowed inside the vehicle. (Patton Museum)*

Above: *In this photograph, an SCR-508 radio set is mounted to the right rear of the cab of the T29. A canvas cover with snaps is fitted over the front of the enclosure to protect the radio. On the right rear of the body is a whip antenna and antenna base. (Patton Museum)*

Below: *The cab of an M3A2 is viewed with the windshield armor lowered. On the manufacturer's plate on the right side of the dashboard, the model designation, M3A1, has been modified with a small plate over the number 1 (presumably with the number 2). (Patton Museum)*

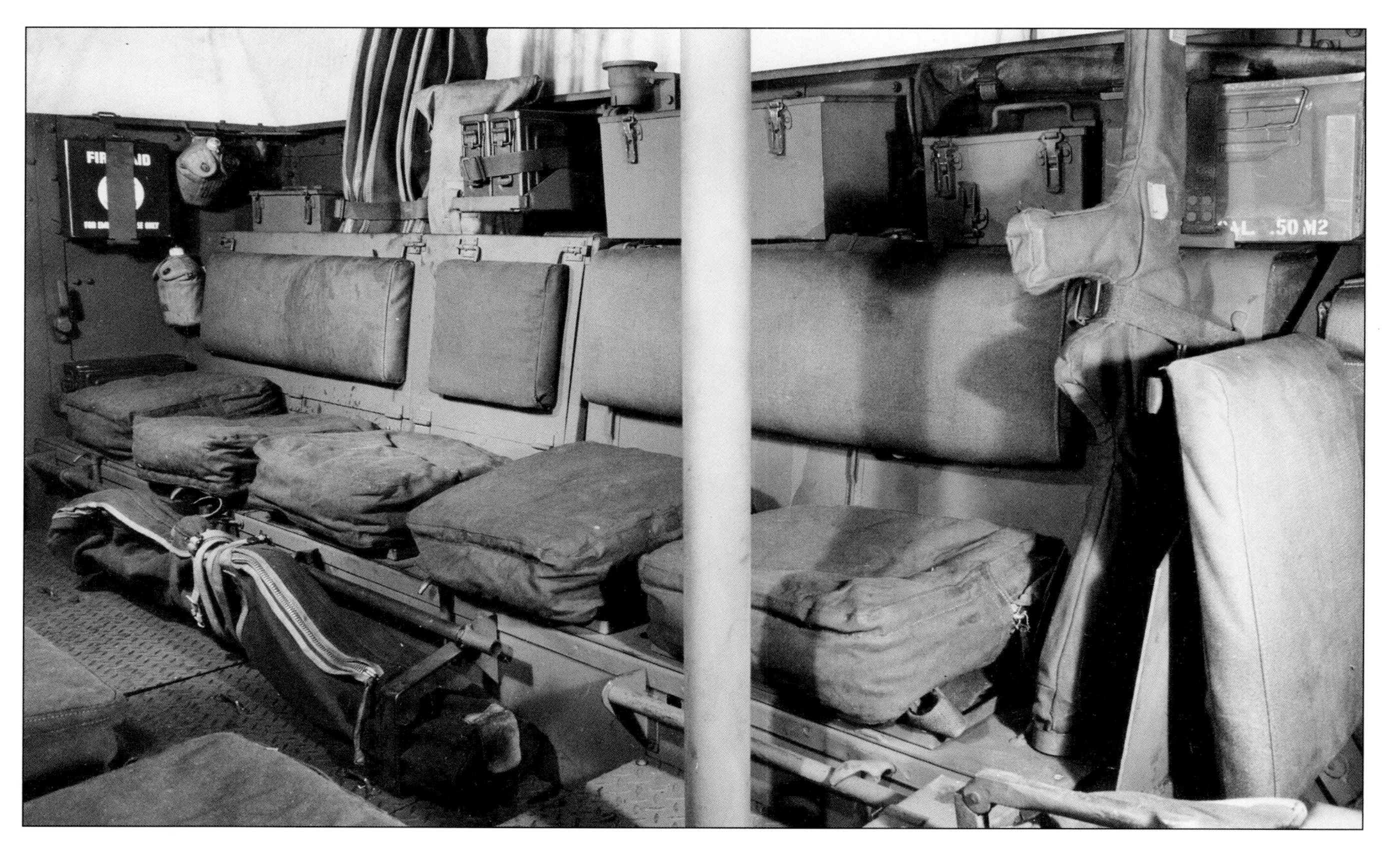

Above: *The rear interior of an M3A2 is observed from the front right passenger's seat in a September 1944 photograph. The layout of the seats, seat backs, first-aid kit, weapons, canteens, and stowage boxes is shown.* **Below:** *The right side of the rear of an M3A2 is displayed. The first several seats are adjacent to the right fuel tank. Folded blankets were stowed inside the seat cushions. To the far left is a fire extinguisher. In the foreground is the ring-mount pedestal. (Patton Museum, both)*

Above: *The left side of the engine compartment of Autocar M3A2 Ordnance number 42977 is shown. Part of the White logo is visible on the side of the engine block to the lower rear of the distributor. To the left is a tag indicating the coolant is rated to -62 degrees Fahrenheit. (Patton Museum)*

Below: *In the right side of the engine compartment are, left to right, the air cleaner, the carburetor and manifold, and the oil cleaner, Fram filter body No. 5267. Two bonding straps for radio-interference suppression are attached to hood-hinge screws. (Patton Museum)*

Chapter 7
The M5 Personnel Carrier

The International Harvester Machine

In the waning days of 1941, the automotive plants of International Harvester had been instructed to find jobs utilizing their facilities in order to offset the decline in truck production. With the Japanese attack on Pearl Harbor, there was a huge demand for military vehicles, including half-tracks—in fact in 1942 the army was forecasting a need for 76,000 half-tracks in 1943 and even more in 1944—and there was little hope that the facilities of the three firms in the Half-track Engineering Committee could keep pace.

To fill this gap, OCM 18050 of 9 April 1942 directed that an additional half-track manufacturing facility be established. In anticipation of this development, Ordnance officials in March 1942 had begun negotiating with International Harvester toward having that firm manufacture half-tracks. Rather than supply plans or blueprints, Ordnance shipped a Diamond T half-track to Harvester's Fort Wayne Works, where it arrived on 20 March.

Incidental to this, advances in welding thin armor plate had been made such that on 16 January 1941 the Ordnance Committee had recommended that two M2 and one M3 body of welded construction be procured. Each of the three half-track manufacturers, Diamond T, White and Autocar, were to submit a body, which would be subjected to comparative firing tests against bolted bodies, using a .30-caliber at 200 yards. The vehicles, designated M2E1 and M3E1, were to be procured as rapidly as possible. Tests at Aberdeen showed that the welded construction provided superior protection as compared to conventional bolted construction; accordingly, welded body construction was specified.

On 28 March 1941, International Harvester's Chicago headquarters directed the company's Fort Wayne Works to prepare estimates on producing half-tracks at the rate of 1,000 vehicles per week. The instructions given were that the estimate was to be based on the body being built by International's Springfield Works, and mounted there.

On 9 April 1942 the Ordnance Committee asked that pilot models of the IH vehicles be procured with the model designations M2E5 and M3E2. When OCM 18370 was approved on 18 June 1942, the designations had been changed to M9 and M5 respectively, and the vehicles had been classified as Substitute Standard. They were designated Standard Nomenclature List (SNL) G-147.

A purchase order was issued to International on 21 April 1942 calling for the production of 7,519 M5 half-tracks. Beyond the body, these vehicles differed from the G-102 (White, Autocar, Diamond T) vehicles in many ways. One of these was the desire to use a more powerful engine, which in turn would permit the use of steel tracks. This situation was contemplated in view of the critical rubber shortage at the time. Of course, this would complicate the supply channel of half-track parts, which had been largely interchangeable. To minimize the impact of this non-uniformity, it was decided that the IH half-tracks would be supplied for international aid, with

Two mechanics make adjustments to engine components on an M5A1 half-track. The radiator fan is propped up against the windshield armor. Above the fan is the pulpit containing an M49 ring mount, which differentiated the M5A1 from the M5 half-track. (Clark County Historical Society)

the half-tracks of the other firms, Standard Nomenclature List class G-102, being retained for U.S. use.

Two Diamond T M3 half-tracks were supplied to Harvester, who, as directed by the Chicago Ordnance District, modified these vehicles to incorporate IH components, including the 450 Red Diamond engine and IH-produced front and rear axles, and one was to be fitted with larger springs. The modified vehicles were designated M3E2. This order was followed quickly by an order for three production pilots, which were to incorporate welded homogeneous armor bodies. Compared to the bolted armored used by the other manufacturer, the welded armor was thicker, and with an attendant weight penalty. In addition to the three pilots ordered by the government, the company built two additional pilots for their own use, one to be used by the Ft. Wayne engineering department and the other for the parts catalog division.

The International Harvester personnel carrier was classified Substitute Standard as the M5 by OCM 18370 on 18 June 1942. Externally, the M5 can be distinguished from the M3 by the lack of screws, the rounded rear corners of the body, and the front fenders formed from flat stock, as opposed to the more elegant truck-type front fenders used by other manufacturers. The front tires of the International half-tracks were 9.00-20, whereas the G-102 vehicles used 8.25-20 tires; this change was incidental to the different drive ratios of the various IH-manufactured components.

Following testing of the pilot model at the General Motors Proving Ground in late 1942, production of IH half-tracks began. While the engineering was done at Fort Wayne, the vehicles were assembled on assembly line C at the Springfield works. M5 deliveries began 21 December 1942. As with the G-102 series vehicles, several changes were made during the production run. In mid-1943 an Issue Letter was released directing the incorporation of "Protectoseal" fuel caps, as had been the case with the G-102. Shortly thereafter, provisions began to be made for track ice scrapers. A 24 July 1943 Issue Letter directed the incorporation of a slave battery terminal outlet in the battery box, allowing for fast and easy jump starting. Luggage racks were added to the rear of the half-track bodies following a 22 September 1943 issue letter.

Modification Work Order (MWO) G147-W4 of 12 November 1943 provided direction to convert a Carrier, Personnel, Half Track M5 to an ambulance through the installation of litter carrier brackets and supports.

When it directed the installation of the M49 ring mount and so-called pulpit, OCM item 20368 of 6 May 1943 brought about the most visible change to the M5. The vehicles so equipped would be designated M5A1. Of the M5s listed below, the last 2,959 vehicles, beginning 2 October 1943, were actually completed as M5A1, featuring a ring mount. International Harvester delivered the last M5A1 in March 1944.

Truscon Steel was awarded contract W-33-019-ORD-659, production order number T-11549, for the remanufacture of 1,799 M5s in during 1944 and 1,408 M5s and 65 M5A1s during 1945.

Just as a "universal" half-track was developed in the G-102 series, a universal half-track combining the capabilities of the M5A1 and M9A1 was developed for the International Harvester chassis. Developed under the authority of OCM 20438 dated 30 April 1943, the new model was initially designated the T31. On 20 July 1943 it was recommended via OCM 21501 that the vehicle be adopted as Substitute Standard and designated M5A2. This recommendation was approved by OCM 21782 on 2 October 1943. While on 23 October 1942 Services of Supply diverting six M5 for developing a universal body for the M5 and M9 had been approved by W.A. Wood, Director, Requirements Division, this was disapproved by the Services of Supply on 6 March 1943, citing the International Aid use of the vehicle. The disapproval was rescinded on 22 April 1943, and three M5s were shipped to Fort Wayne for this purpose.

Above: *In January 1941, the Ordnance Department placed an order for three half-tracks with welded bodies. Two were to be based on the M2 half-track car and one on the M3 half-track personnel carrier. White produced these pilot vehicles. Shown here is one of two examples based on the M2, designated the M2E1. (ATHS)*

Below: *The use of screws on the M2E1 was kept to a minimum, but they were used in the fabrication of the diamond-tread steps and the bumperette assemblies on the rear of the vehicle. The M2E1 has homogeneous steel armor, as opposed to face-hardened steel on the M2. (ATHS)*

The driver's area of a standard M2 half-track car. The reinforcing plate between two body panels (top left) and the vertical strip to the rear of the door are attached to the body with screws and nuts. (ATHS)

On the M2E1, the reinforcing plate and the vertical strip are welded to the body. Tests of the M2E1 at Aberdeen Proving Ground demonstrated that the welded construction reduced lead splashing and ricocheting screws and nuts when hit by projectiles. However, the homogeneous armor was more readily penetrated than the M2's face-hardened steel. (ATHS)

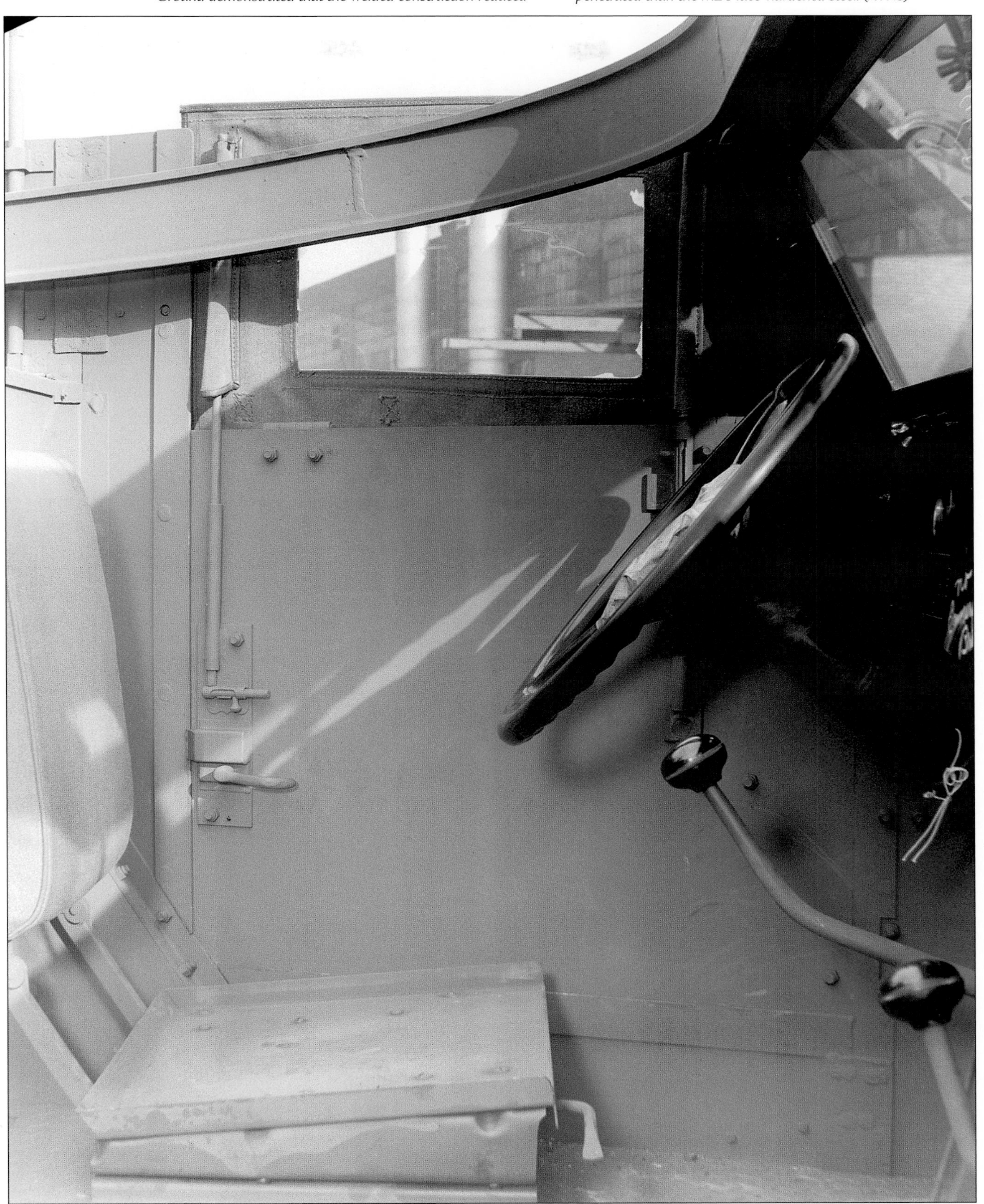

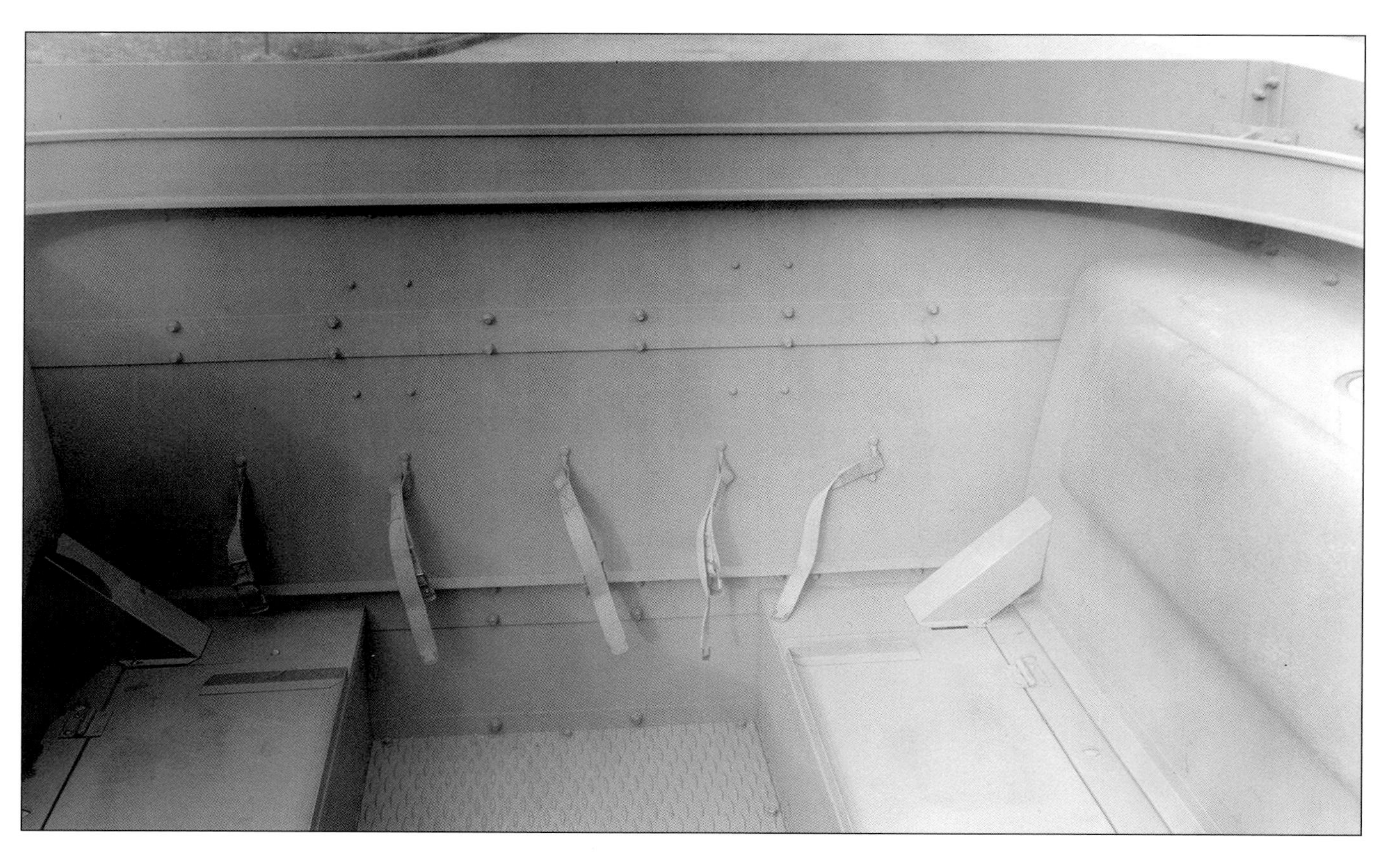

Above: *The section of skate rail to the rear of the interior of the body of the M2 half-track is fastened to the rear armor panel with 16 screws arranged in vertical pairs.* **Below:** *The skate rail's mounting brackets are welded to the upper body, rather than attached with screws through the upper body. Footman loops and webbing straps were provided for stowing ammunition boxes. Note the retainers for holding the seat cushions in place. (ATHS, both)*

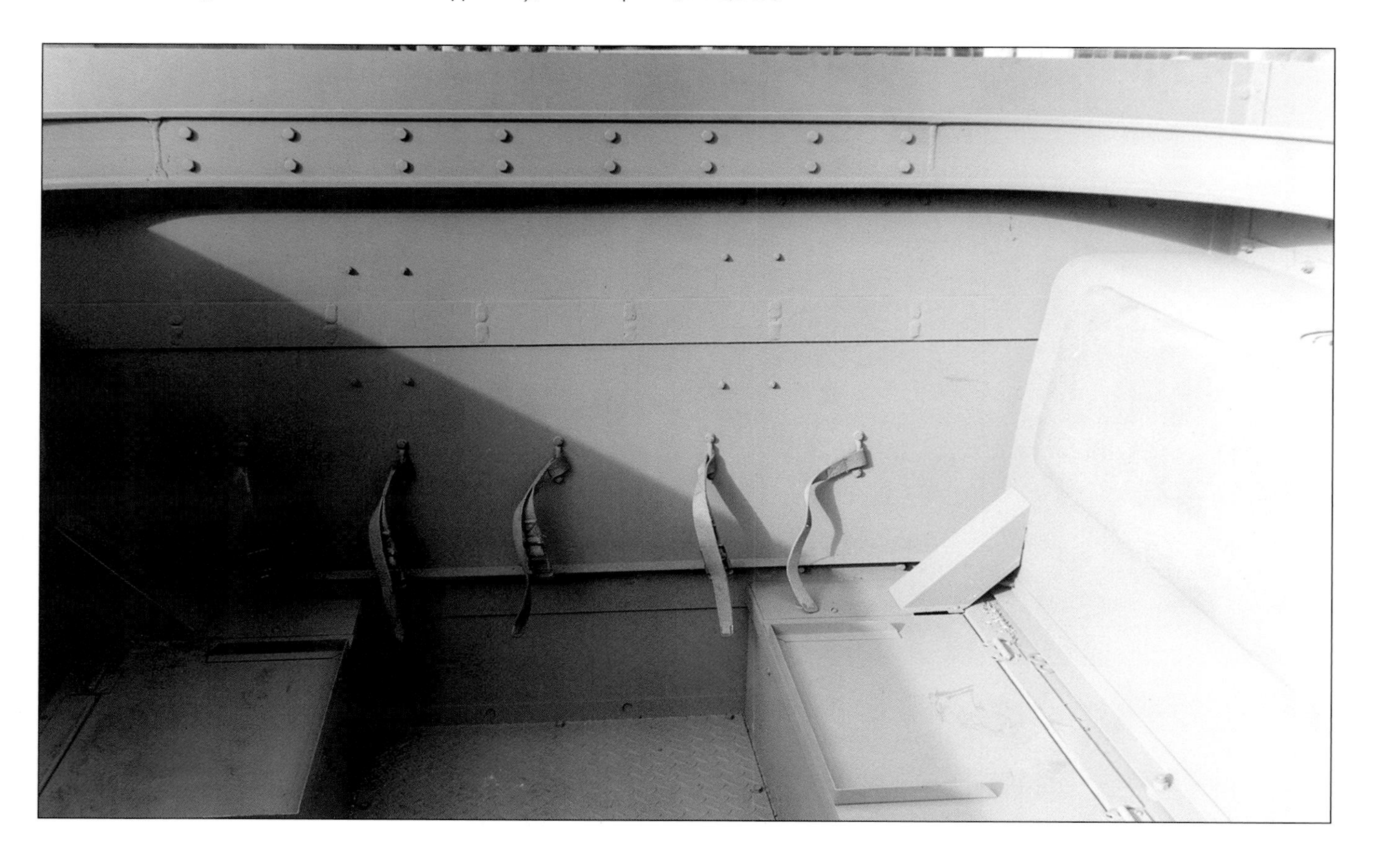

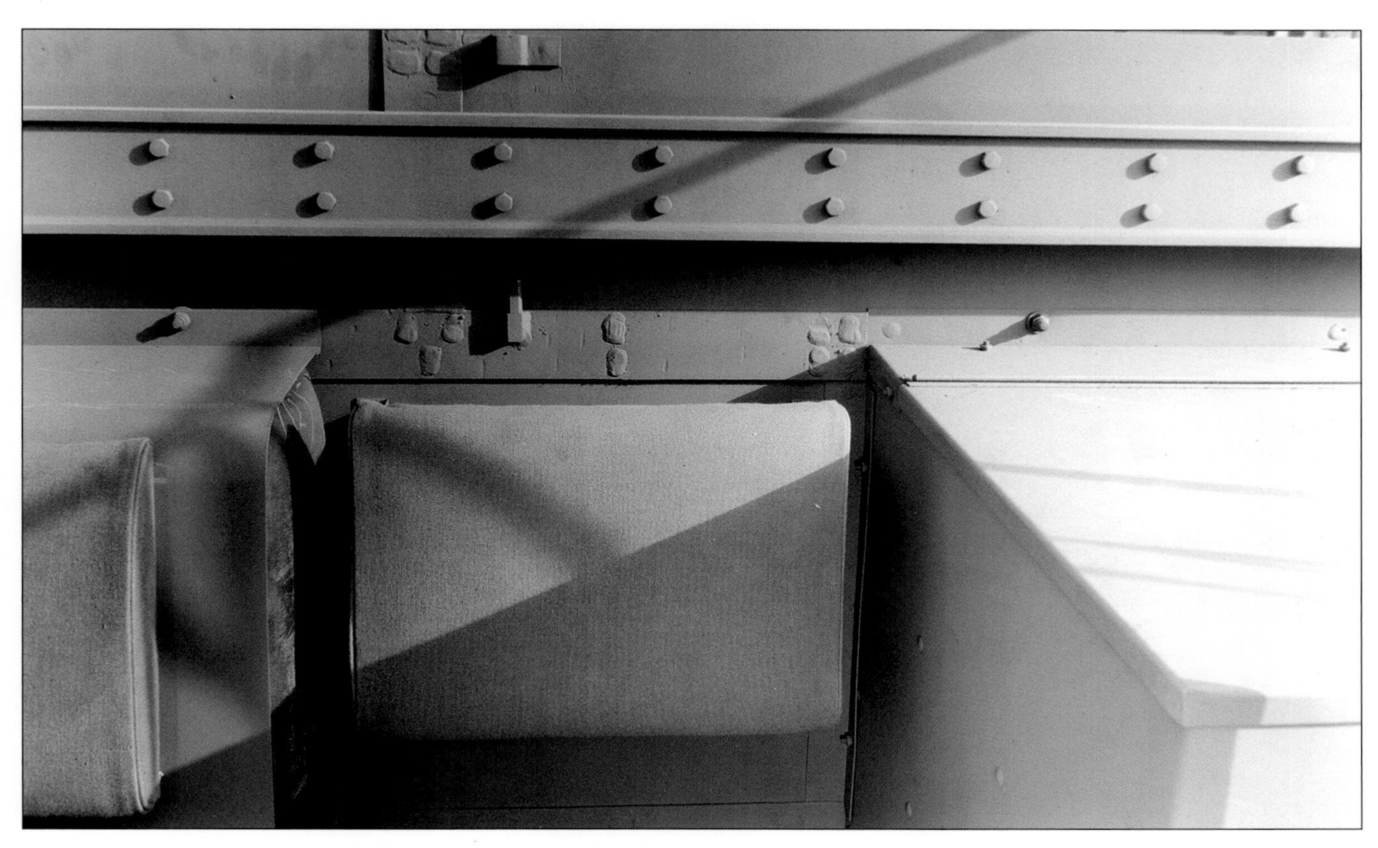

Above: *The seat between an M2E1's fuel tank (left) and ammunition locker (right). (ATHS)* **Below:** *Because of the inability of White, Autocar, and Diamond T to keep up with America's mushrooming demand for half-tracks after the country's entrance into World War II, in early 1942 the Ordnance Committee ordered pilot vehicles of two new types of welded-body half-tracks from International Harvester Company. These were the M2E5, similar to the M2, and the M3E2, the counterpart of the M3. The M3E2 was standardized as the M5 half-track personnel carrier in June 1942. (TACOM LCMC History Office)*

Above: *Work is underway on a G-147 half-track chassis at International Harvester. The men are working on the engine and radiator, while in the foreground the right bogie assembly with the track-support rollers is lying on top of the bogie frame.* **Below:** *The worker to the right is installing a pintle-hook assembly into the rear of the chassis frame of an International half-track. The continuous rubber tracks have been installed, and the man to the left is tightening the nuts retaining the outer idler flange. (Clark County Historical Society, both)*

Above: *Photographed at the Detroit Arsenal in December 1942, the M5 is in some ways externally similar to the late-model M3, with detachable headlights, liquid container racks, and mine racks. There are two major external differences: the fenders are flatter in cross-section than those of the M2/M3 half-tracks and lack their rolled edges and fairing at the rear; and the rear corners where the sides of the body meet the rear are rounded. (TACOM LCMC History Office)*

Below: *The body of the M5 is fabricated from welded rolled homogeneous steel, 5/16-inch thick. The windshield cover and the sliding visor plates on the door are 5/8-inch thick. Spring-loaded idlers are standard equipment for the M5s. The outer edges of the fenders have a beveled appearance. Powering the M5 was the IHC Model RED-450-B 6-cylinder engine through the Spicer Model 1856 transmission. (TACOM LCMC History Office)*

Like the M3 half-track personnel carrier, the M5 had seating for thirteen, stowage spaces and rifle racks behind the rear seats, and a pedestal for a machine gun toward the front of the crew compartment. A structural frame runs along the upper perimeter of the body, wrapping around to the edges of the rear door. International Harvester delivered 4,625 M5s from December 1942 to September 1943. (ATHS)

A worker wearing a face mask uses a grinder to smooth rough edges on the header on which the windshield armor is hinged. Part numbers are stenciled on some components: for example, 132564-11/C on the vision-port covers. (Clark County Historical Society)

Above: *A technician at International Harvester tests the gauges and controls on an instrument panel destined for an M5-type half-track. Each instrument panel was thoroughly inspected before it was installed in the vehicle.* **Below:** *Engine assemblies are being prepared for installation in International Harvester M5 half-tracks. On the two closest assemblies, details of the flywheel housings, transmissions, transfer cases, flanges, and universal joints are visible. (Clark County Historical Society)*

An IHC RED-450-B engine assembly is being hoisted in preparation for installing it on an M5 half-track chassis. Underneath the manifolds is the carburetor. Below the hoisted engine is another engine assembly with the transmission, hand brake, PTO, and transfer-case control levers. (Clark County Historical Society)

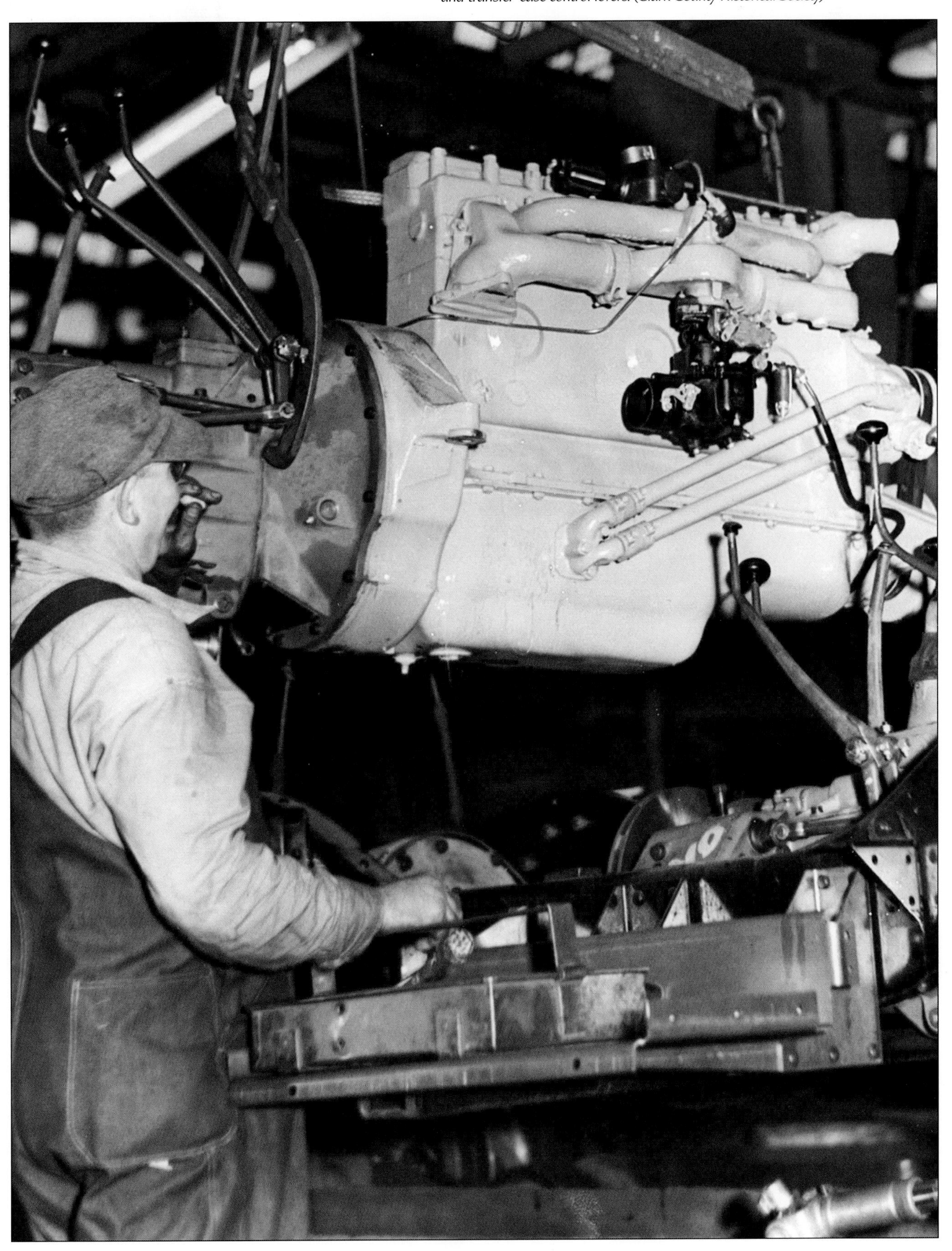

Elements of a right-hand M5 bogie assembly and, to the right, sprocket are shown close-up in this assembly-line photo. Notice the prominent molding seams around the center of the rubber bogie-wheel tires. (Clark County Historical Society)

Above: *Workmen using crowbars wrestle a track assembly onto the right-hand running gear of an M5 half-track. To fit the track in place, the bogie wheels have been compressed and held in place with a chain. (Clark County Historical Society)*

Below: *Under assembly here is the part of the body to the rear of the cab of an International half-track. The plates being installed are those to the rear of the cab. (Clark County Historical Society)*

As a half-track body moves along the IH Springfield assembly line, workers are installing the oil bath air cleaner on the right forward side of the cowl. At far right in this photo can be seen the voltage regulator. Along the firewall is visible part of the wiring harness as well as the carburetor choke cable. (Clark County Historical Society)

Above: *The worker in the ring mount in this M5A1 is mounting the carriage for the machine gun. Two fixed socket mounts for .30-caliber machine guns are visible on the top rails of the half-track body.* **Below:** *International Harvester employees pose for a photo in the last International half-track, an M5A1 covered with graffiti. Such "artwork" sometimes was applied to special vehicles, such as the 10,000th example to be completed, or, as in this case, the last one to leave the assembly line. (Clark County Historical Society, both)*

The rounded rear corners of the body of the M5 half-track are evident in this view from above, as are the distinctive bumperettes, somewhat similar in shape to a Jeep's. The seat cushions are not installed. As in the M3 half-track, the fuel tanks are placed to the rear of the front seats. Note the diagonal braces for the seat backs to the rear of the fuel tanks. (ATHS)

Above: *The addition of the M49 ring mount to the M5 half-track resulted in a new model, the M5A1. International Harvester manufactured 2,959 M5A1s from October 1943 to May 1944. This example was photographed at the Studebaker Proving Ground on 8 November 1943. Note the canvas curtain installed over the door. (TACOM LCMC History Office)* **Below:** *An M5A1 poses for a documentation photo at the Ordnance Operation, Engineering Standards Vehicle Laboratory in Detroit on 29 April 1944. The tarpaulin and a cover for the ring-mounted .50-caliber machine gun are installed. On the right fender is a large, rolled camouflage net. The full complement of twelve antitank mines is in the mine rack. (TACOM LCMC History Office)*

Above: *The same M5A1 is shown with the tarpaulin removed. Note the rounded rear corner of the body and the absence of screws on the armor plates. M1 rifles and M1 carbines are poised in the racks in the crew compartment. The Browning M2HB .50-caliber machine gun is situated in a Model D40733 cradle-and-pintle assembly on the M49 ring mount. (TACOM LCMC History Office)* **Below:** *The rear of the M5A1 (as well as the M5) is easily distinguishable from the M3 and M3A1 half-track personnel carriers by the rounded corner of the bodywork and the bumperettes. Poking from the right rear of the tarpaulin is the covered barrel of the .30-caliber machine gun. (TACOM LCMC History Office)*

Left: *On the M5A1, the upper body panels protrude slightly with reference to the lower panels, on which the taillights, towing pintle, and bumperettes are mounted. The receptacle for the trailer lights and brakes connection is crowded in closely to the left taillight. A canvas bag holding a tripod for the .30-caliber machine gun is strapped to the rear door. (TACOM LCMC History Office)*

Below: *The tarpaulin cover has been folded and strapped to the left fender of this M5A1, which is under evaluation at the Engineering Standards Vehicle Laboratory in April 1944. Strapped to the top of the tarpaulin is a tripod for the .50-caliber machine gun. Stenciled on the body to the rear of the driver's door are "PREPARED BY L.T.D. 3/11/44" and "NO RADIO," although a radio antenna is present. (TACOM LCMC History Office)*

Above: *Some details of the inner side of the M49 ring mount are visible. Footman loops are fastened to the body and upper door panels for securing the tarpaulin. A thumbscrew on the sliding armored panel in the visor on the upper door panel locked the panel in the desired position. The sliding panels on the windshield cover worked in the same manner. (TACOM LCMC History Office)* **Below:** *This M5A1 half-track personnel carrier was photographed at the Detroit Arsenal. The canvas tarpaulin was of fairly complex construction, with multiple panels sewn together, including reinforcing strips. This tarpaulin lacks the side flaps sometimes found on other U.S. half-track tarpaulins. To the upper rear of the driver's door is an extension in the tarpaulin to accommodate the radio antenna mount. (TACOM LCMC History Office)*

Above: *An overhead view of an M5A1 at the Engineering Standards Vehicle Laboratory in April 1944 shows the locations of the three socket mounts for the .30-caliber machine gun: two on each side of the crew compartment and one at the right rear. The design of the M49 ring mount and how it fits within the armored semi-enclosure is also apparent. (TACOM LCMC History Office)* **Below:** *The outer edges of the sheet-metal fenders of the M5A1 are bent to provide added strength, and have a beveled appearance. Goodyear "All Service" nondirectional-tread tires are mounted on this vehicle. The standard tire size for the M5 and M5A1 half-tracks was 9.00-20. (TACOM LCMC History Office)*

Above: *The shutters on the M5A1 are rolled homogenous steel, 5/16-inch thick. They can be opened or closed using a control lever under the right side of the dash board. The brush guards for the headlights are of a noticeably different shape than the several types found on the M2/M2A1 and M3/M3A1 and their derivatives. (TACOM LCMC History Office)*

Below: *In the right side of the engine compartment of an M5A1 half-track are a Donaldson Model 9241 oil-bath air cleaner and the International Harvester RED-450-B engine. The intake manifold is mounted directly above the exhaust manifold. Below the center of the exhaust manifold is the carburetor. (TACOM LCMC History Office)*

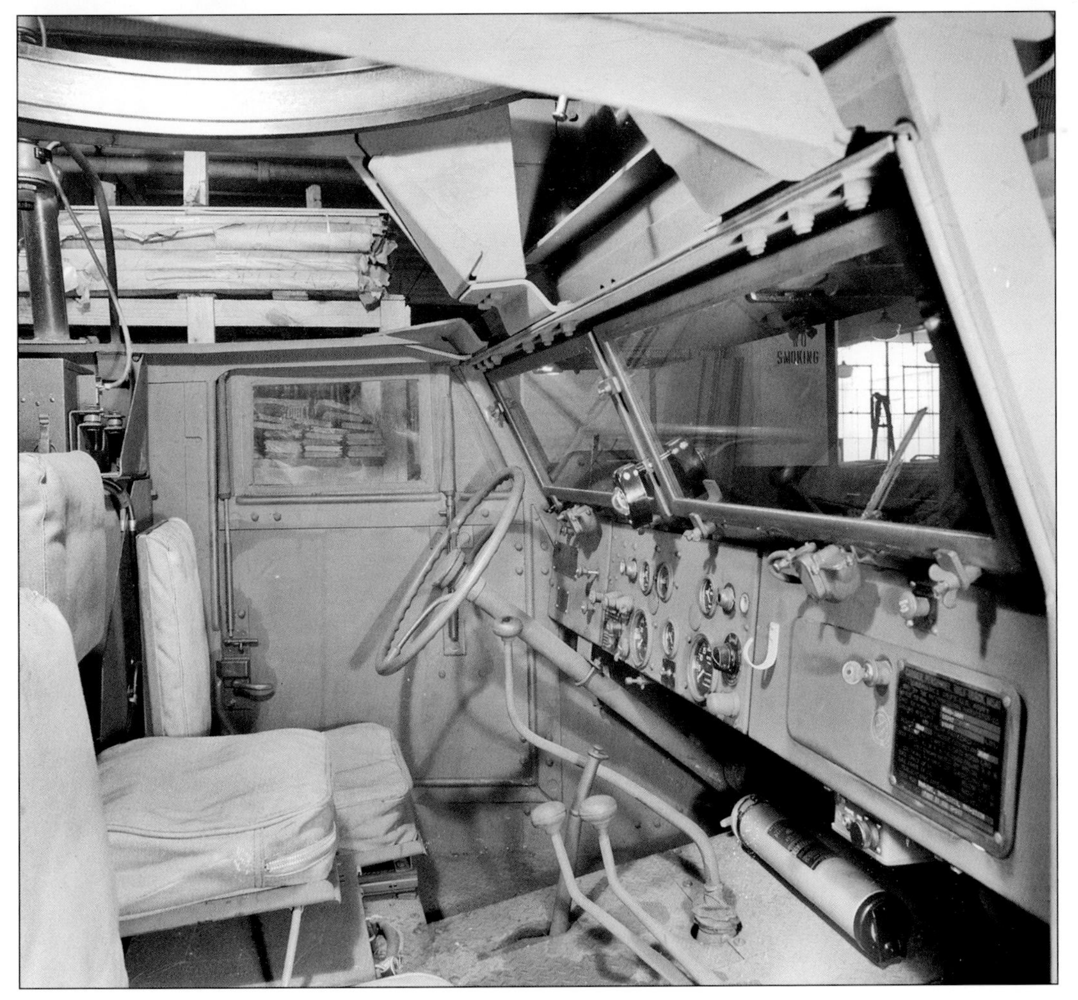

Above: *On the left side of the M5A1's engine are, left to right, the belt-driven generator (Delco-Remy Model 1117308), distributor (Delco-Remy Model 1110161), two oil filters, and surge tank. Cast on the engine crankcase is "RED DIAMOND." The black box next to the surge tank is the Delco-Remy generator regulator. (TACOM LCMC History Office)*

Below: *A view through the right door of an M5A1 photographed in February 1944 displays the structure of the two front supports of the M49 ring mount, toward the upper right. The canvas curtain with clear plastic window is attached to the driver's door. To the front of the transmission gearshift lever is a decontamination apparatus, and below the glove box is an intercom box. (TACOM LCMC History Office)*

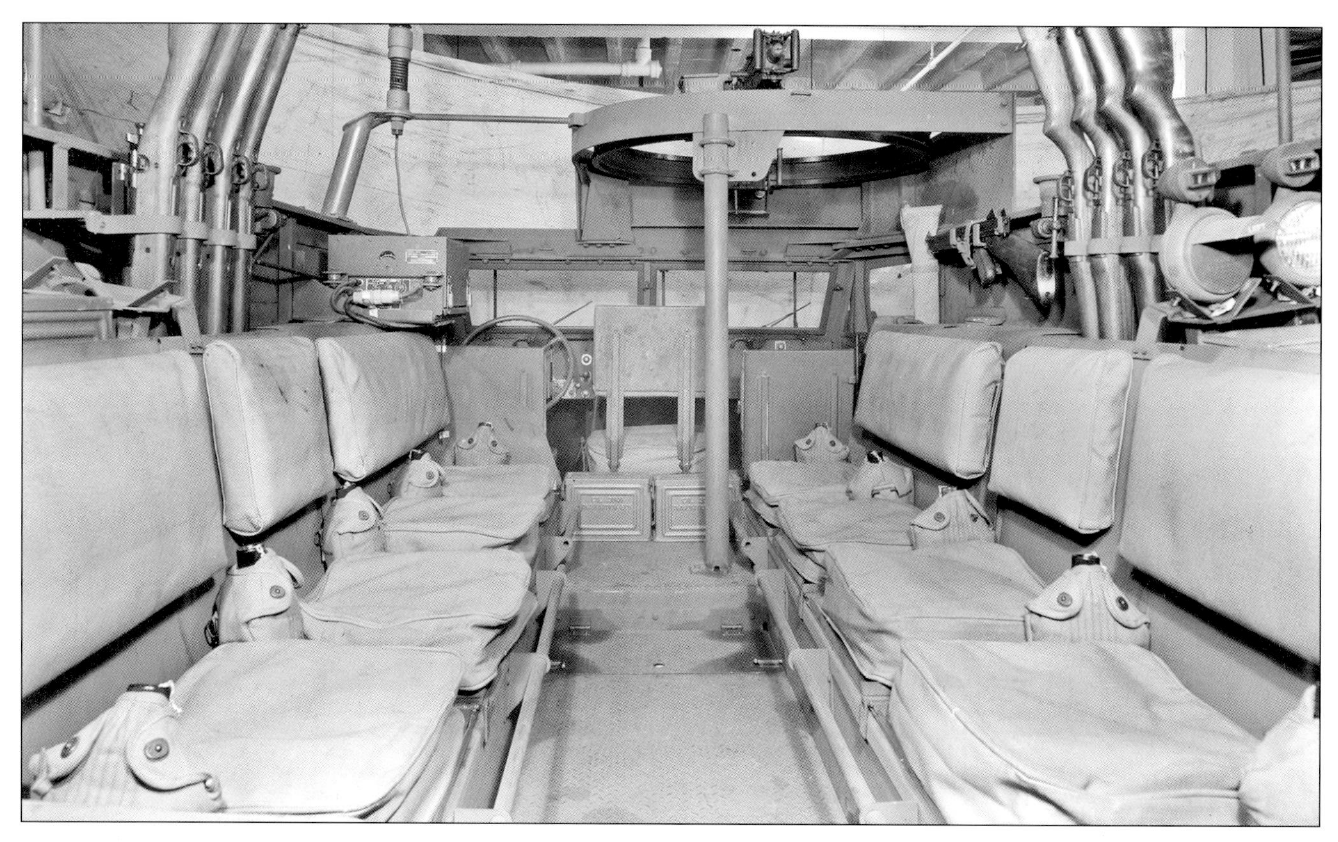

Above: *The crew compartment of an M5A1, as viewed from the rear. On the forward right side is a radio set with the antenna rising above it. To the front center is the rear support post for the M49 ring mount. On the right wall of the vehicle is a .45-caliber Thompson machine gun and, to the far right, spare service and blackout headlights. Two .30-caliber ammunition boxes are to the rear of the center front seat. (TACOM LCMC History Office)* **Below:** *In a view of the same M5A1 facing toward the rear, the ring mount is to the lower left. A close-up view is offered of the socket mount to the far right, with its clamp handle at the bottom. Below the removable floor plates is stowage space for vehicular tools and equipment. A mix of M1 Garand rifles and the by-then obsolete M1903 Springfield rifles are stored in the racks. (TACOM LCMC History Office)*

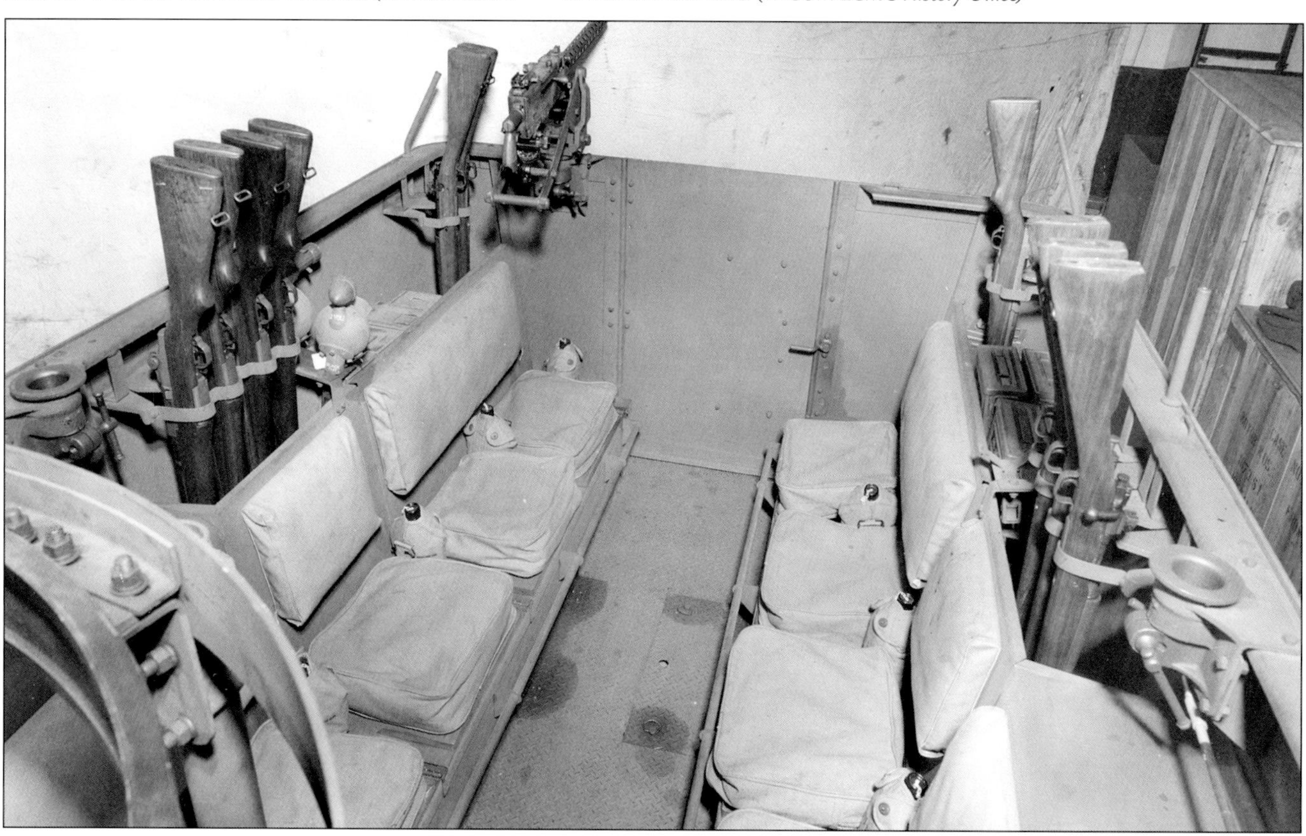

Above: *This file photo of a winch-equipped M5A1 was taken for the Development and Engineering Department of the Detroit Arsenal on 25 June 1943. A 6-foot ruler has been propped against the bumper for scale. Note the side braces on the outboard sides of the headlight brush guards. (Patton Museum)*

Below: *On the M5A1 (and other models of half-tracks with ring mounts), a separate panel is fitted over an opening in the main tarpaulin over the ring of the gun mount, allowing the gunner to operate the machine gun without removing the entire tarpaulin. In addition, there is a cover for the .50-caliber machine gun itself. (TACOM LCMC History Office)*

Above: *The crossed straps on the front of the canvas cover over the ring mount of this M5A1 secure the front of the cover to the main tarpaulin. Below the passenger's door is the battery box, containing a six-cell, lead-acid-type battery rated at 168 amperes at the six-hour rate. The battery was quite heavy, weighing 165 pounds. (TACOM LCMC History Office)* **Below:** *Several M5A1 half-tracks with full canvas, including the door curtains with soft plastic windows, are lined up. The U.S. Army registration number of the closest vehicle appears to have been 4069075. (Clark County Historical Society)*

Visible U.S.A. numbers on these M5A1 half-tracks include 4069094 on the nearest vehicle, a winch-equipped example, and 4069177 on the third one, which has a front roller. The unique brush guards for the headlights are also seen to good advantage. (Clark County Historical Society)

Left: *In a photo taken for the Chief of Ordnance, Detroit, an M5A1, registration number W-4034216, undergoes winter testing to evaluate its ability to operate on snowy, uneven terrain. The .50-caliber machine gun has been removed from the ring mount, and the canvas cover for the ring mount is in position but bunched up. The radiator shutters are closed. (TACOM LCMC History Office)*

Below: *This M5, assigned to the 93rd Armored Recon Battalion, has been outfitted for use by unit mechanics. An air compressor was mounted at the right rear of the body, across from a workbench with a vise. The unit, assigned to the 13th Armored Division, trained at Camp Beale, CA, and Camp Bowie, TX before landing at Le Harve in January 1945. (Patton Museum)*

Above: *The T31 represented the pinnacle of G-147 development. Described as the Universal half-track, the vehicle combined the attributes of the M5A1 and M9A1 into a single vehicle. (Patton Museum)*

Below: *Although the T31 was standardized as the M5A2 in October 1943, the vehicle, like the similar M3A2, never entered series production. The body layout accommodated both personnel carrier and command vehicle requirements, along with the weapons-mountings of a reconnaissance vehicle. (Patton Museum)*

Chapter 8 The M9A1 Car, Half-track

More International Aid from International Harvester

Even as negotiations were ongoing concerning the vehicle that came to be known as the M5, the Ordnance Department on 11 July 1942 asked International Harvester to submit a proposal for their version of the M2. Like the IH-built M5, this vehicle was also to be produced expressly for international aid. By August the discussions had progressed to the point that the army was requesting that production of 2,974 of the vehicles begin in February 1943—yet a design had not yet been established.

Two White-produced M2s had been delivered to International on April 20, and had been modified with IH components in the same manner as the M3E2 described in the previous chapter. The modified M2s were designated M2E5. However, in addition to an M2E5-type vehicle, International also proposed a vehicle with the same body as size as the M5—that is, approximately 10-inches longer than the M2 body, and also incorporating the rear door. This larger body was the style that the Chicago Ordnance District opted for on 11 September 1942. On 26 October the previous request for 2,974 vehicles was amended to 4,659 vehicles and 46 ½ sets of spare parts. Half of the vehicles were to be equipped with unditching rollers and the remainder with front-mounted winches—increasing the cost of each by about $187.00.

On 10 October 1942 a contract was issued for two pilot models of the vehicle that would be the M9. Before this vehicle could be built, on 31 March 1943, the Chicago Ordnance District requested that the ring mount be used rather than the skate rail, and the designation of vehicles so equipped be changed to M9A1. Thus, when the pilot was delivered on 17 March 1943, it was in fact an M9A1. On 17 April Ordnance requested a price proposal for an additional 3,771 M9A1 vehicles. Deliveries of production vehicles began on 30 April 1943 when 10 M9A1 vehicles were shipped.

Although on 15 July 1943 a notice of award had been issued to IH for the production of a total of 8,327 M9A1, on 15 September 1943 the Chicago Ordnance District issued a partial notice of termination, reducing the quantity to 4,535. This was followed by a second notice in December that cancelled a further 1,102 of the vehicles, and ended the M9 program. International Harvester delivered the last M9A1 in December 1943.

U.S. Lend-Lease records located thus far lump the M5, M9 and M9A1 vehicles into a single category of "M5 series." Deliveries, excluding theater transfers, are listed as: Mexico, 2; Chile, 10; Canada, 20; USSR, 420; British Empire, 5,238. British records list the following IH half-tracks:

Z5305008-5306507, contract SM6105; total 1500
Z5517050-5518172, contract SM6173; total 1123
Z5541298-5541689, contract SM6176; total 392
Z5580022-5580771, contract SM6176; total 750
Z5820073-5820472, contract SM6176; total 400
Z6110575-6111574, contract SM6176; total 1000
Z6155025-6155524, contract SM6176; total 500
Z6098821-6101127, contract SM6176; total 2,307*

*Described as "reconditioned," so may have been ex-U.S. Army.

Above: *Manufactured by International Harvester (IHC) from March to December 1943, the M9A1 half-track is similar in design and purpose to the M2A1 half-track car and features the M49 ring mount. This example, registration number 4068670, photographed in August 1943, displays typical M9A1 features, including the lack of side doors for the ammunition lockers, IH-style fenders and headlight brush guards, and a welded body. (TACOM LCMC History Office)*

Below: *A rear-left, three-quarters view shows another M9A1 with the canvas tarpaulin removed. The suspension is essentially identical to that of the M2 and M3 families of half-tracks. (TACOM LCMC History Office)*

U.S.A.
4068670

A technical manual photo displays the layout of the rear of the M9A1 half-track. Behind the front seats on both sides of the crew compartment are ammunition lockers, accessible only from the inside of the vehicle. Fuel tanks are at the rear corners of the crew compartment. The bumperettes are the same as those used on the M5 and M5A1 half-tracks. (Wisconsin Historical Society)

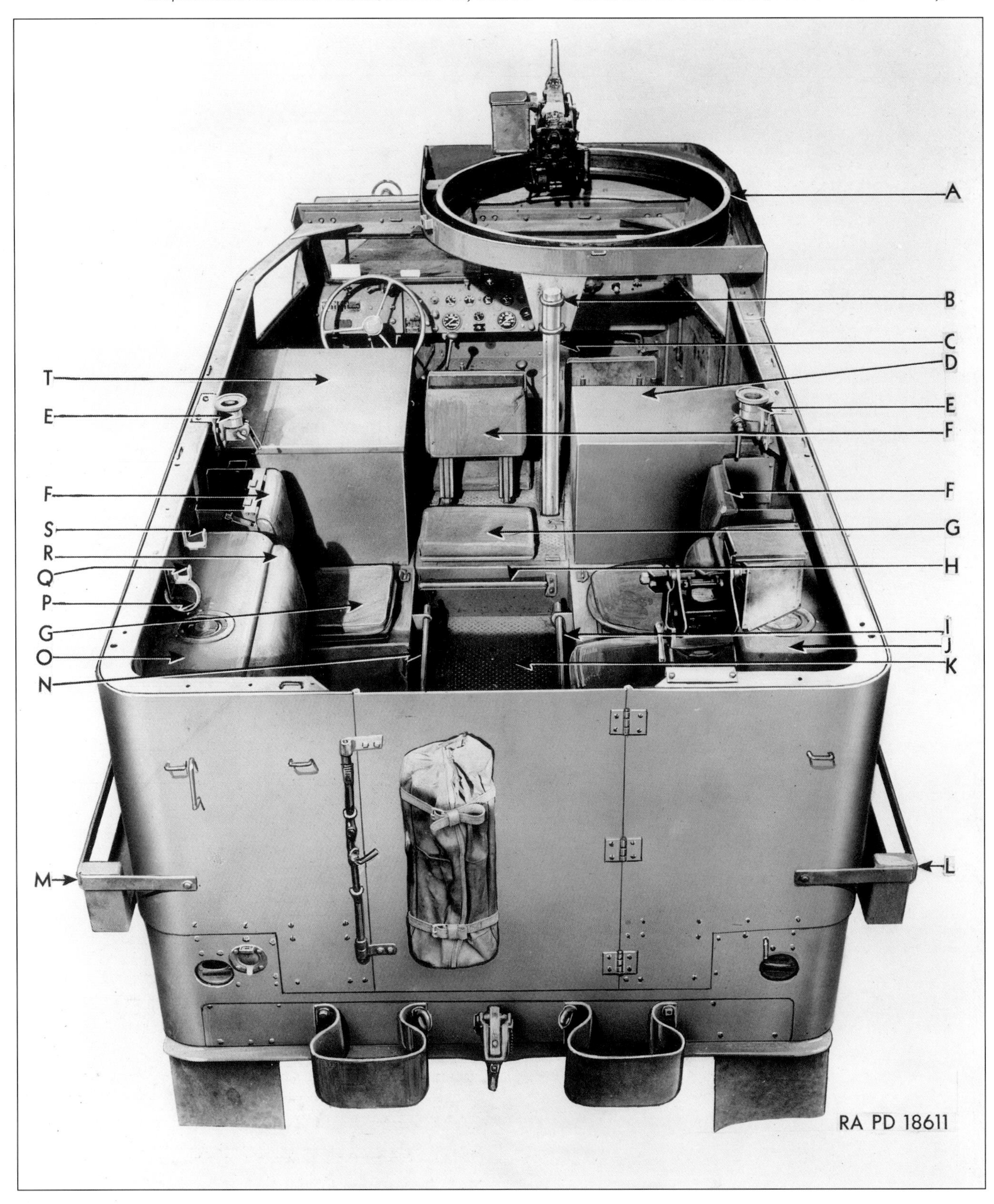

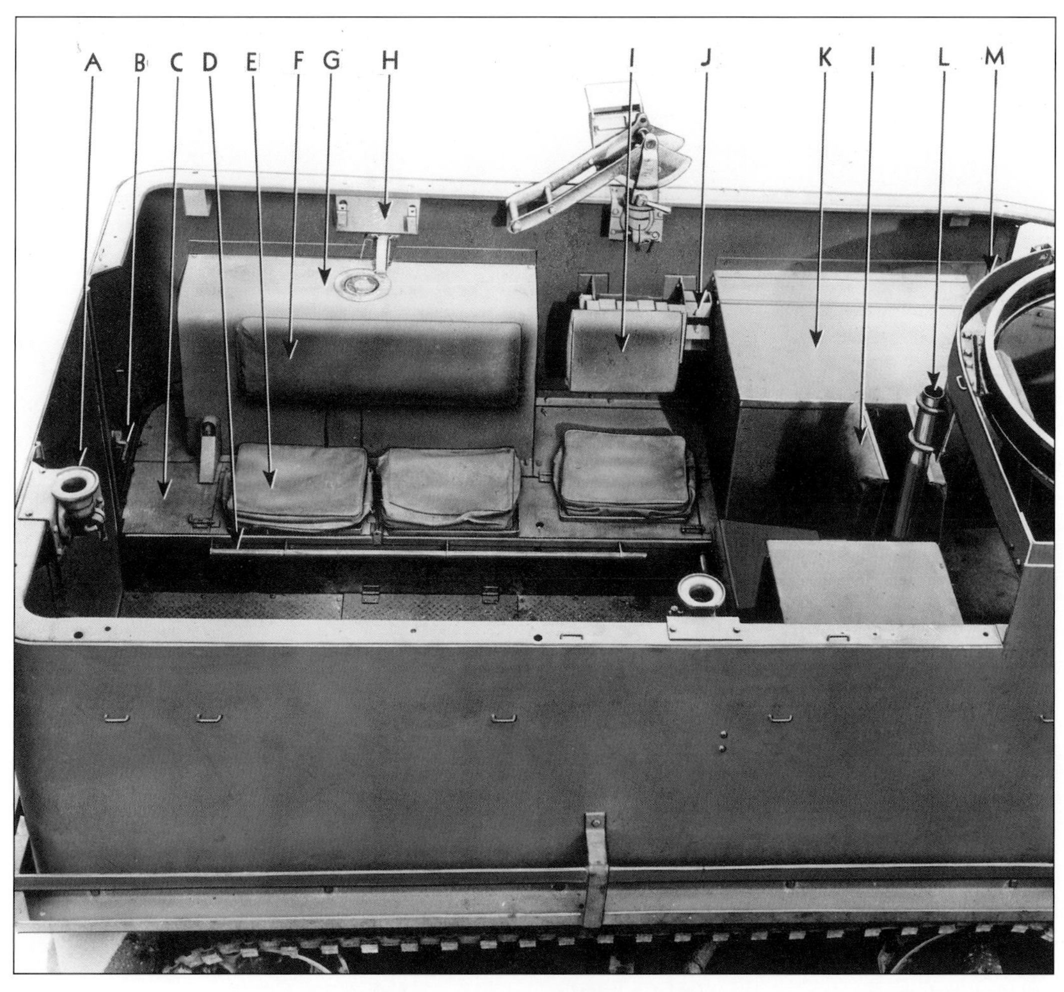

Above: *This technical manual photo shows the left side of the crew compartment of an M9A1. Two seats are next to the fuel tank, and a single seat is positioned between the fuel tank and the ammunition locker. Above that single seat is a socket mount for a .30-caliber machine gun; the cradle and pintle assembly for that gun is shown installed in the socket, and another socket is next to the rear door. (Wisconsin Historical Society)*

Below: *This M9A1, registration number 4070166, was photographed at the Detroit Office of the Chief of Ordnance, Engineering Standards Vehicle Laboratory, on 11 February 1944. The IH-style fender and the mounting bracket on the underside of the fender are visible. This vehicle mounted the Tulsa 10,000-pound winch in front. (TACOM LCMC History Office)*

Above: *The nondirectional 9.00-20 tires were characteristic of late-war half-tracks. The stencils on the side of the body read "PREPARED BY L.T.D. 11/18/43," "SCR-193," and "SCR-528," the latter two being references to two radio sets installed in this vehicle. The antitank mines bear stencils identifying them as "INERT LOADED." (TACOM LCMC History Office)*

Below: *With the tarpaulin cover removed from the same half-track, the two radio antenna mounts are visible. The tarp is rolled up and strapped to the left fender, and a camouflage net is stowed on the right fender. Note the details of the underside of the fender and its supports. (TACOM LCMC History Office)*

Above: *The M2/M2A1 and M3/M3A1 half-tracks are easily distinguishable from each other when viewed from the rear, both because of the noticeably different arrangement of their bumpers and by the presence or absence of a rear door. However, the M9A1 (shown here) and the M5A1 are virtually indistinguishable from each other from the rear. (TACOM LCMC History Office)*

Below: *The same M9A1 as in the preceding photo is shown here from the right rear with its tarpaulin cover removed. The smooth surface of the welded body armor is evident, with only a few screws, the footman loops for the cover tarpaulin straps, and other necessary pieces of hardware intruding on the otherwise uncluttered plates. (TACOM LCMC History Office)*

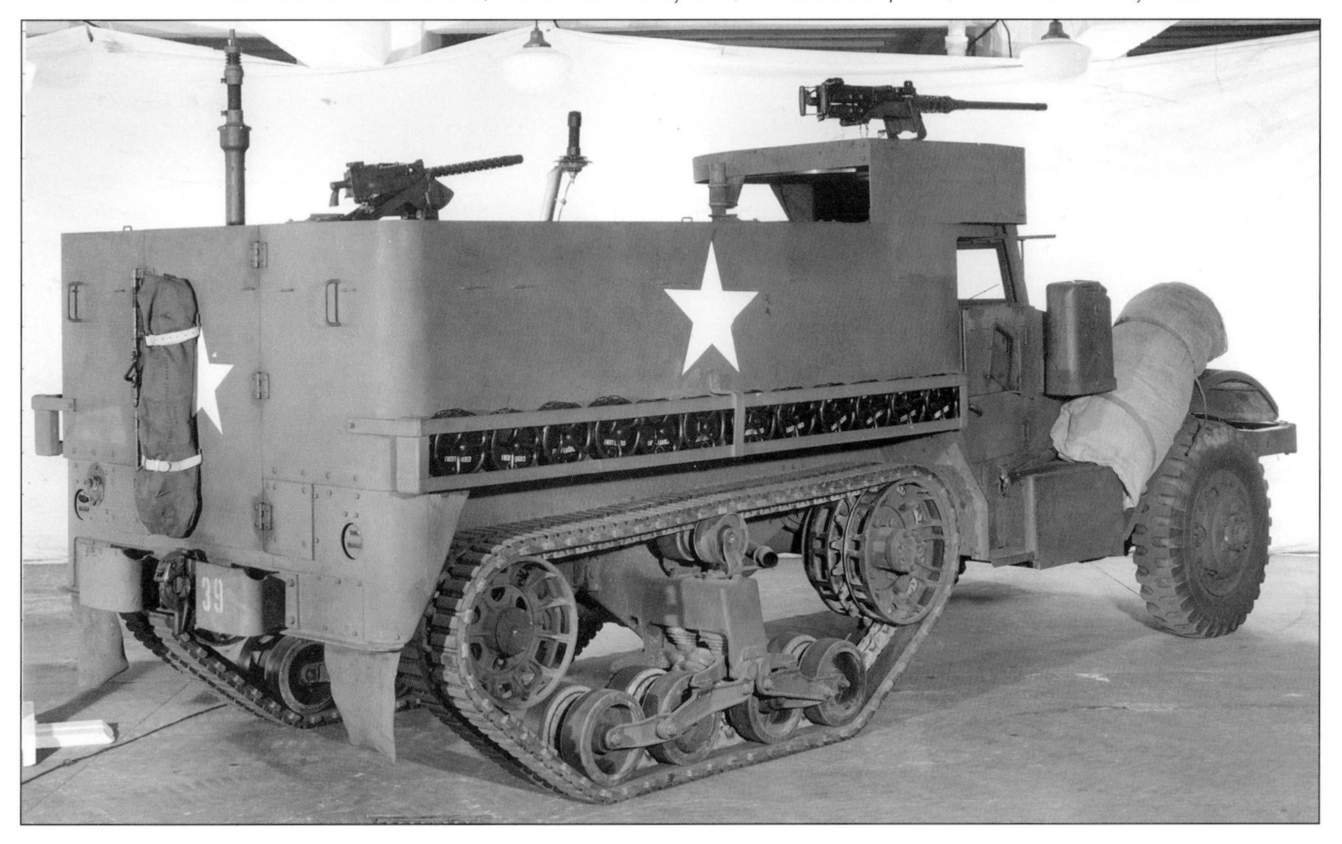

Above: *This view of the left side of the M9A1 under evaluation at the Engineering Standards Vehicle Laboratory, on 10 February 1944, emphasizes the relative locations of the .50-caliber and .30-caliber machine guns and antenna mounts. On IH half-tracks, the sliding side-vision shutters and their brackets are on the inside of the upper door panels, whereas on the M2 and M3 families of half-tracks, the shutters and brackets are on the outside. (TACOM LCMC History Office)*

Left: *The radiator shutters of this M9A1 are closed, displaying the recognition star painted on them. The front axle assembly is the International Harvester Model FDK-1370, which weighs 820 pounds and is a full-floating, single-reduction type with a conventional differential. The design of the headlight mounts is also evident in this photo. (TACOM LCMC History Office)*

Above: *This M9A1, viewed from the right side, displays the armored side protection for the M49 ring mount. Above the tire, the front support of the fender is visible. The tailpipe of the exhaust is to the front of the bogie frame. The idler wheels are equipped with the late-type coil springs. (TACOM LCMC History Office)*

Left: *The latch handle and link rods for the rear door are mounted on the outside of the M9A1, to the left of the door. An additional handle on the inside of the crew compartment allows the door to be unlatched from within. Stenciled in black on the tripod cover on the door is "COVER TRIPOD MOUNT OVERALL" and the part number, D71865. (TACOM LCMC History Office)*

Above: *This M9A1 under evaluation at the Engineering Standards Vehicle Laboratory, Detroit, was photographed from above on 2 March 1944. The arrangements of the three socket mounts around the crew compartment and the extra radio in front of the rear door are evident. The cover for the .50-caliber machine gun on the ring mount is stenciled with the part number D40726, along with nomenclature information. (TACOM LCMC History Office)*

Below: *The cab, as viewed from the right door. Note the hinges at the front of the frame of the passenger's seat and on the back-rest frame. These allow the back rest to fold down and the seat to fold forward. On the door are the support bolt for the canvas side curtain and a holder for the lubrication chart. (TACOM LCMC History Office)*

Above: *As viewed from outside of the rear of the M9A1 at the Engineering Standards Vehicle Laboratory, facing forward, the top of the housing of the rear radio set is at the bottom. Another radio, under cover, is to the left behind the driver's seat. The left ammunition locker has been removed to make room for the latter radio, but the right ammunition locker is still in place. (TACOM LCMC History Office)*

Below: *As depicted in an M9A1 technical manual illustration, the passenger's (or squad/section commander's) seat back is folded down. To the rear of the center front seat is a rear-facing seat. Beyond that seat is the left ammunition locker. (Wisconsin Historical Society)*

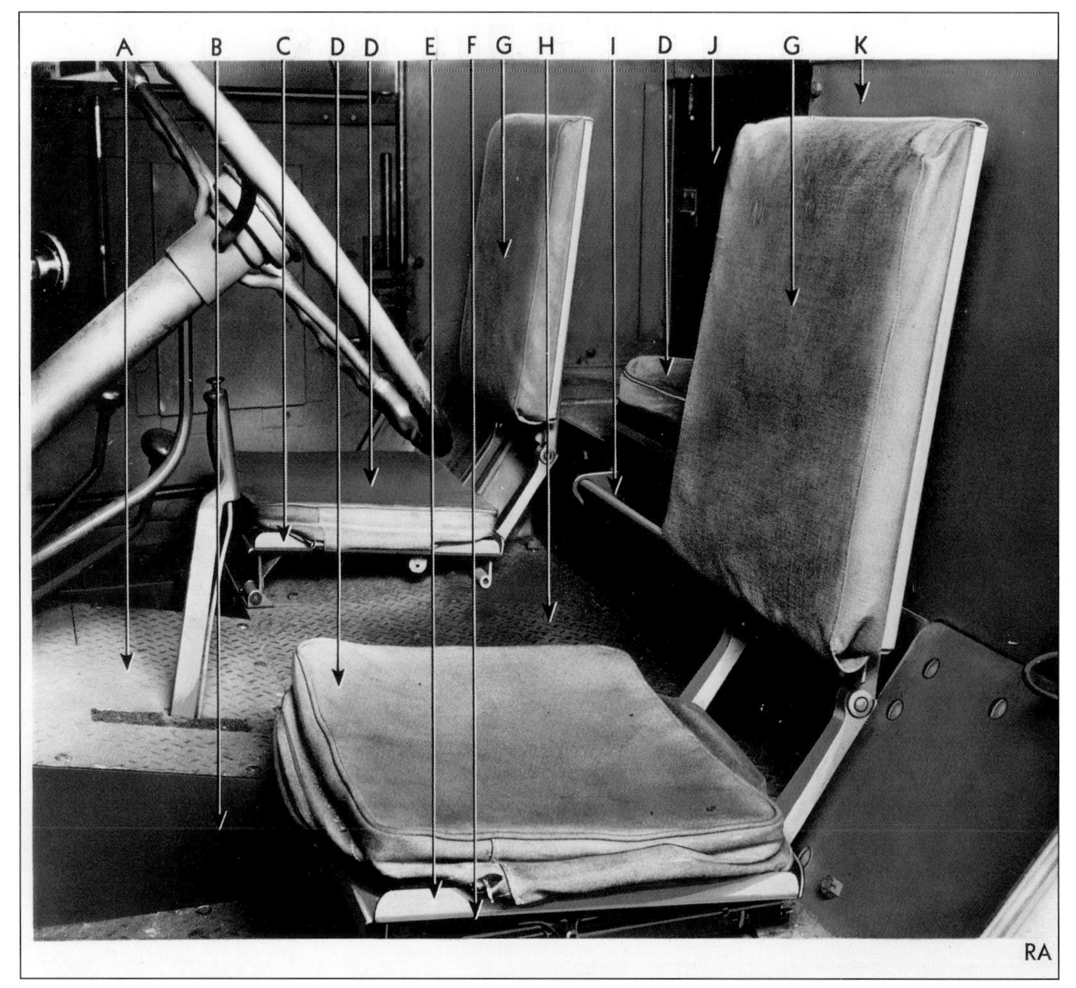

Above: *Another technical manual photo shows the driver's and section commander's seats from the left side of an M9A1, with a small part of the front center seat showing. Also in view are the steering wheel, hand brake, and shifting levers for the transmission and transfer case. Folded blankets could be stuffed into the seat cushions, but have been not inserted in this instance. (Wisconsin Historical Society)*

Below: *The arrangement of the engine compartment of the M9A1 half-track is generally identical to those of the M5 and M5A1 half-tracks. Shown from the right side of the compartment is the International Harvester RED-450-B engine, with the oil-bath air cleaner to the left. On the fold line of the hood, above the front of the valve cover, is a braided bond. This is part of the radio suppression equipment. (TACOM LCMC History Office)*

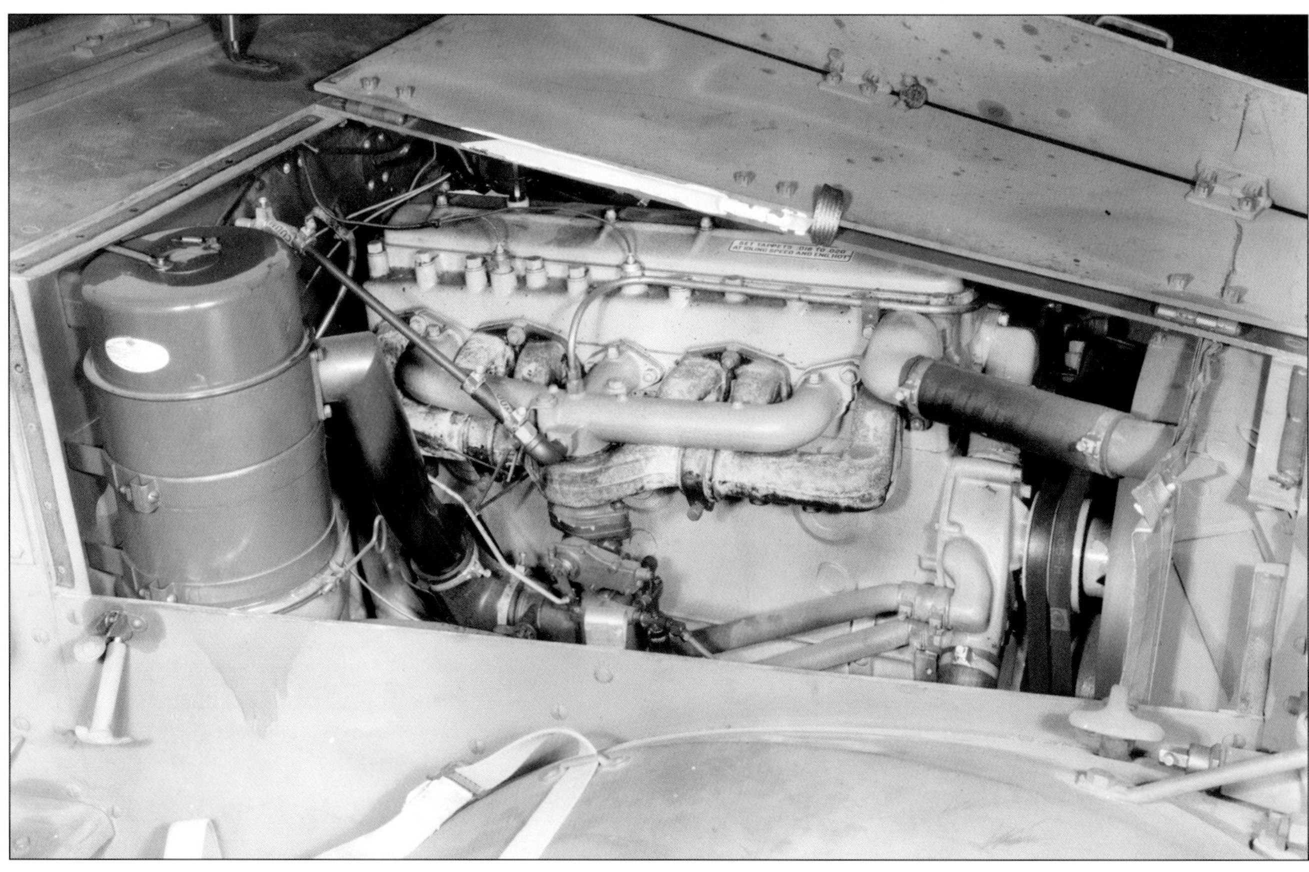

Above: *The left side of the M9A1 engine compartment includes, left to right, the Delco-Remy Model 1117308 generator, two oil filters, and Delco-Remy Model 1110161 distributor. Below the surge tank, toward the lower right, is a sticker providing information on the engine lubricant and additives added at the factory, and instructions on the type of fuel to use during the break-in period. (TACOM LCMC History Office)*

Below: *In this photo of an M9A1, the absence of the vision shutter and its bracket on the upper panel of the side door is evident. The interior shutter and bracket is an identifying feature of International Harvester half-tracks. Although a model of International Harvester half-tracks designated M9 was authorized, these were all given M49 ring mounts at the factory and entered service with the designation M9A1. (Patton Museum)*

Above: *This M9A1, registration number 4067760, was photographed during evaluation at the Developing and Engineering Department, Detroit Arsenal, on 25 June 1943. "M9A1" and "177" are stenciled in white on the side door. (TACOM LCMC History Office)* **Below:** *An M9A1 half-track with no visible registration number is parked outside of a factory building. Visible above the ring mount is a machine-gun cradle and ammunition box. (The Patton Museum)*

Above: *The rear ends of the mine racks of International Harvester half-tracks are different in design than those of the M2 and M3 families of half-tracks. On IHC mine racks, such as the M9A1 shown here, the frame at the top is bent at the rear and is attached to the rear of the body.* **Below:** *A technical manual illustration shows an M9A1 from the right rear. Attached to the left side of the machine gun cradle-and-pintle assembly on the rear socket mount is an ammunition tray with beveled bottom. This tray has a hinged cover. (Wisconsin Historical Society, both)*

Above: *The rounded rear corners of the body of this half-track, combined with the M49 machine gun mount and the fuel tanks located in the rear of the crew compartment, indicate that it as an M9A1. To the left of the photo, one ammunition locker is present. On the opposite side, radio equipment is installed. Writing on the photo identifies this as the half-track of the commander of Recon Company, 93rd Reconnaissance Battalion. (Patton Museum)* **Below:** *Two G.I.s scrub down an M5 or M9-series half-track, distinguishable by its distinctive headlight brush guards. The half-track parked next to the one in the foreground, with the curved rear corner of the body, is also an International Harvester vehicle. In the background, another half-track bears a crossed cannons insignia of the artillery branch. (TACOM LCMC History Office)*

Above: *In preparation for the upcoming invasion, Auxiliary Territorial Service women and their civilian assistants work on M5A1 or M9A1 half-tracks at a vehicle reserve depot in Worcestershire on 25 April 1944. (IWM)*

Below: *A Universal carrier with wading screens attached moves past a line of British M5 half-tracks. All are passing through Hermanville-sur-Mer on the afternoon of 6 June 1944. (IWM)*

Above: *During Operation Epsom on 29 June 1944, British Tommies of the 8th Battalion, Rifle Brigade, 11th Armoured Division, take a break while digging in next to their International Harvester M9A1 or M5A1. The edge of the fender has the distinctive beveled appearance of an IHC half-track. Miscellaneous gear is stowed in the mine rack. (IWM)* **Below:** *An M5 half-track believed to have been in Canadian service proceeds alongside a stone wall. A whip antenna is to the right of the cowl, and unusually tall tarpaulin bows are installed to allow higher headroom inside the vehicle. (The Patton Museum)*

Above: *This M5 half-track with a Canadian registration number has two windows on the side of the tarpaulin. The tarpaulin is fastened to the body with a rather intricate lashing. Camouflage netting is stowed above the cab. (The Patton Museum)*
Below: *A Signals M5 half-track of the British 11th Armoured Division is stuck in the mud in Holland on 19 October 1944, and gets a push from troops. On the left fender is a picture of a bull, the insignia of that division. German-type jerrycans were frequently seen on British half-tracks. The tarpaulin cover has been modified with a different side panel, with what appear to be two clear plastic windows. (IWM)*

Above: *An International Harvester half-track, either an M9 or M9A1, tows an Ordnance QF 17-pounder antitank gun on the approach to the Foglia River during the Gothic Line Campaign in Italy on 1 September 1944. Both the vehicle and gun belonged to the British 269 Battery, 87th Anti-tank Regiment. The vehicle is missing its right taillight assembly. (IWM)* **Below:** *Universal Carriers and M5 half-tracks of the Grenadier Guards race across the open ground in the direction of Nijmegen, Holland in September 1944. (Image Bank of WW2)*

Above: *In another scene from Holland on 18 September 1944, British troops enter the town of Graves to link up with American paratroopers, who had liberated the town the day before. (NARA)*

Below: *An M5 half-track of the 1st Battalion, Grenadier Guards, Guards Armoured Division advances towards the Nijmegen bridges on September 19, 1944. (Image Bank WW2)*

A French-owned International Harvester half-track pauses in Strasbourg in late November 1944 and a local boy poses by the vehicle. Displayed on the front of half-track are souvenirs: a framed photograph of Hitler and a German officer's hat hanging on the left headlight brush guard. (ATHS)

Above: *On 31 March 1945, in the final weeks of the war, British troops in Bocholt, Germany, indulge in some singing next to their M9A1 half-track. The body of the half-track has been modified with hinged upper panels and improvised panels with jerrycan racks at the rear. Hitched to the half-track is an Ordnance QF 17-pounder anti-tank gun. (IWM)*

Below: *Several liberated slave laborers help soldiers from the British 1st Rifle Brigade clean their M5A1 or M9A1 half-track on 26 April 1945. The man on the right uses a machete to chip caked-on mud off the fender. Note the crates stored on the windshield cover and the jerrycans stashed in the mine rack. (IWM)*

Left: *Half-tracks and scout cars halt on a road near Fürstenau, Germany as a nearby factory burns on April 9, 1945. Fürstenau is located in Lower Saxony, in northeastern Germany. Combat in this area was part of the general British drive into northern Germany from their positions in the Netherlands. (IWM)*

Below: *Churchill tanks of 6th Guards Tank Brigade with infantry, carriers and half-tracks of 15th (Scottish) Division wait to advance towards the River Elbe on April 13, 1945. The forward half-track has a large wooden structure topped with a tarp. (IWM)*

Above: *Prime Minister Winston Churchill, accompanied by Field Marshals Bernard Montgomery and Alan Brooke, inspect men and tanks of the 7th Armoured Division "Desert Rats" from an International Harvester half-track during the British victory parade in Berlin, on 21 July 1945. A light-colored handrail has been installed around the crew compartment for the VIPs. Several similarly equipped half-tracks follow. (IWM)* **Below:** *Crewmen, including two wearing fezzes with tassels and leather jerkins, stand in front of their half-tracks, which are either M9A1s or M5A1s. Stripes are painted on the grille and the pulpit of the fourth vehicle, possibly indicating a command vehicle. (The Patton Museum)*

Chapter 9 Unarmed Oddities

The Half-track in Atypical Roles

That the G-102 and G-147 half-track chassis went on to be used as the basis for an array of gun motor carriages and fighting vehicles is well known. Those vehicles are presented in considerable detail in volume two of this book. Less well-known are the variations of the base vehicles, the M2, M3 and M5, shown in the following pages, which include ambulance, ammunition carrier, radio carrier and aircraft refueler.

Work to use the M3 half-track as an armored evacuation vehicle to remove wounded personnel was begun by the Armored Force in 1941. By 20 March 1942 the Armored Force Board had generated a report on an M3 equipped with attachments for carrying four litters. The attachments, consisting of three brackets and six chains, could be installed without welding or drilling. That report concluded with the recommendation that such a kit be adopted as Standard, and "be provided for all half-track personnel carriers, M3, used by the medical detachments of the Armored Divisions." Further, it recommended "That immediate steps be taken to procure sufficient sets of these attachments locally to equip the required number of halftrack personnel carriers for the 1st, 2nd, 3rd, and 4th Armored Divisions."

On 23 November 1942 International Harvester was asked to quote on 1,000 similar kits to convert M5 personnel carriers into ambulances. A further revision of the kit was made by the Armored Medical Research Laboratory, providing additional strength. Tests by the Armored Board pointed to the desirability of removing all stowage boxes and brackets to the rear of the fuel tanks in order to provide greater aisle width, allowing better care from the attendant. The TOE shown in the appendix of this volume indicates that such ambulance vehicles were widely used.

The Tank Destroyer Board in 1944 conducted a series of tests regarding the transporting of 3-inch antitank ammunition in and on the M3, successfully loading 32 rounds in addition to the usual ten men.

Arguably the most unusual looking versions of the M3 were a limited number of unarmored versions produced by Autocar. Three were built as airfield refuelers, while four additional trucks were built utilizing the 9-foot wooden cargo bed designed for the CCKW-352. All of these vehicles were bereft of the armored cab and hood, instead utilizing sheet metal cabs and hoods based on that of the Autocar C-70 truck. These vehicles were intended for the transportation of the SCR-299 radio system, and attendant HO-17 shelter, over rough terrain. Remarkably, those developing the vehicle were unaware that the HO-17 required a 12-foot cargo bed. As a result, the overhang of the shelter precluded the effective coupling of the K-52 trailer that housed the generator to supply the powerful radio system. As a result, the Desert Warfare Center assessment of the vehicle included the rather terse statement "the vehicle is unsuitable as a carrier for the SCR-299 radio set."

This photograph of a U.S. Army Medical Department M3 half-track converted to an armored ambulance was taken at Fort Sill, Oklahoma, in 1943 and was intended to document changes in the seating and stowage arrangements. (NARA)

186724

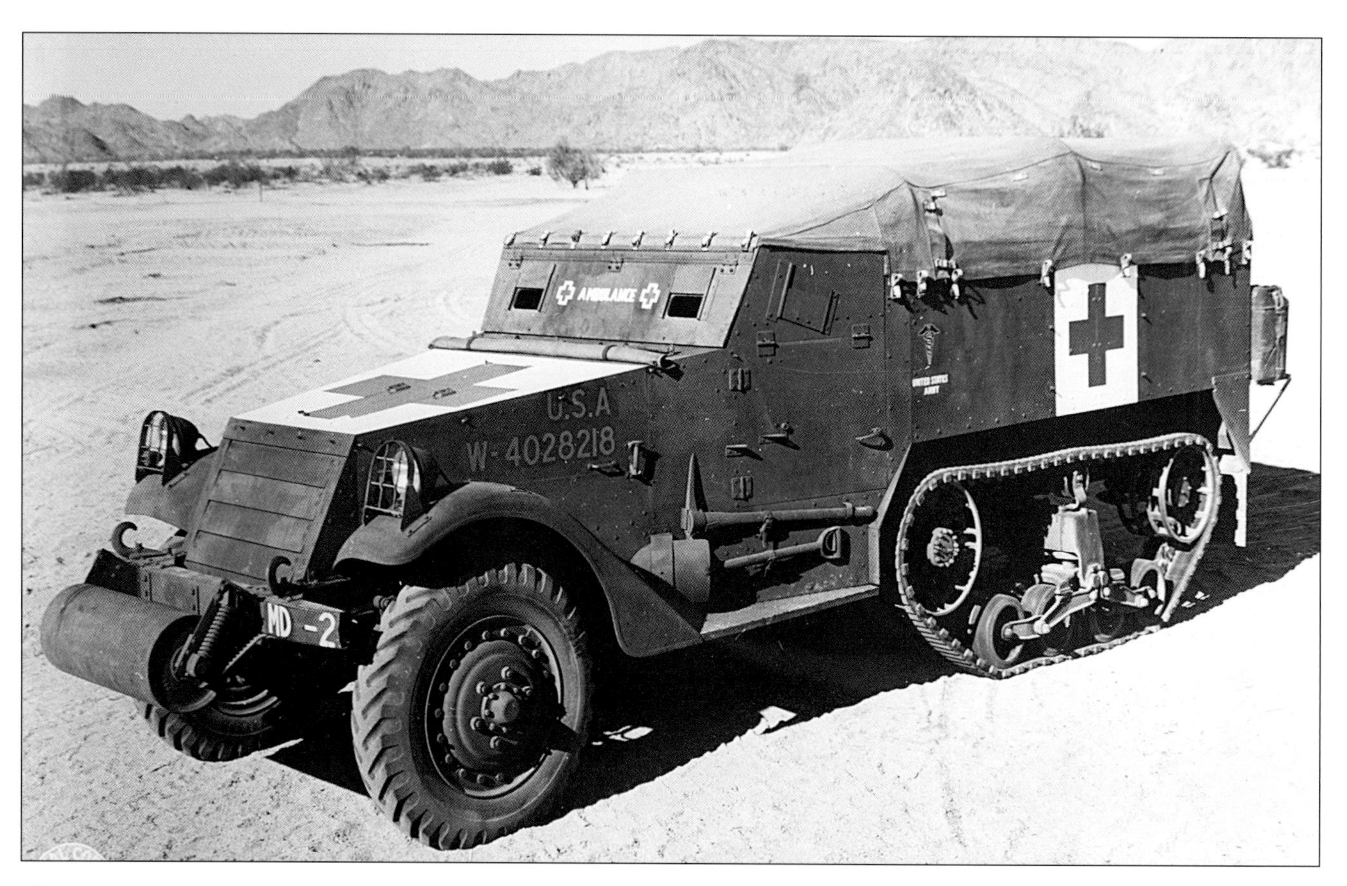

Above: *An M3 half-track converted to use as an armored ambulance was photographed at the Desert Training Center. This vehicle was U.S. Army registration number 4028218. The entire top of the hood was painted white with the Red Cross superimposed.*

Below: *The same M3 half-track ambulance is viewed from the left side. To the rear of the driver's door is the insignia of the U.S. Army Medical Department. On the rear of the body are racks containing 5-gallon liquid containers. (NARA, both)*

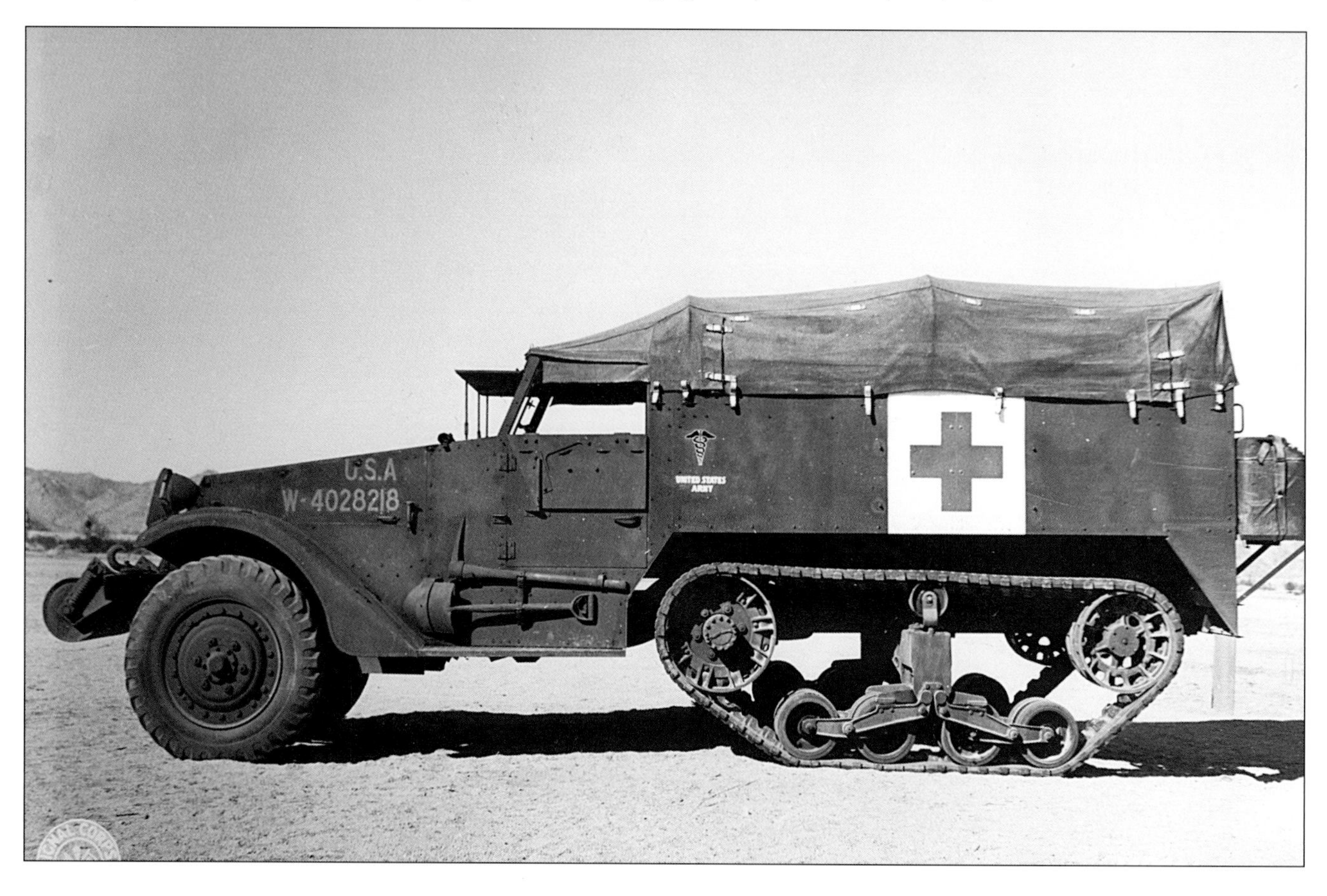

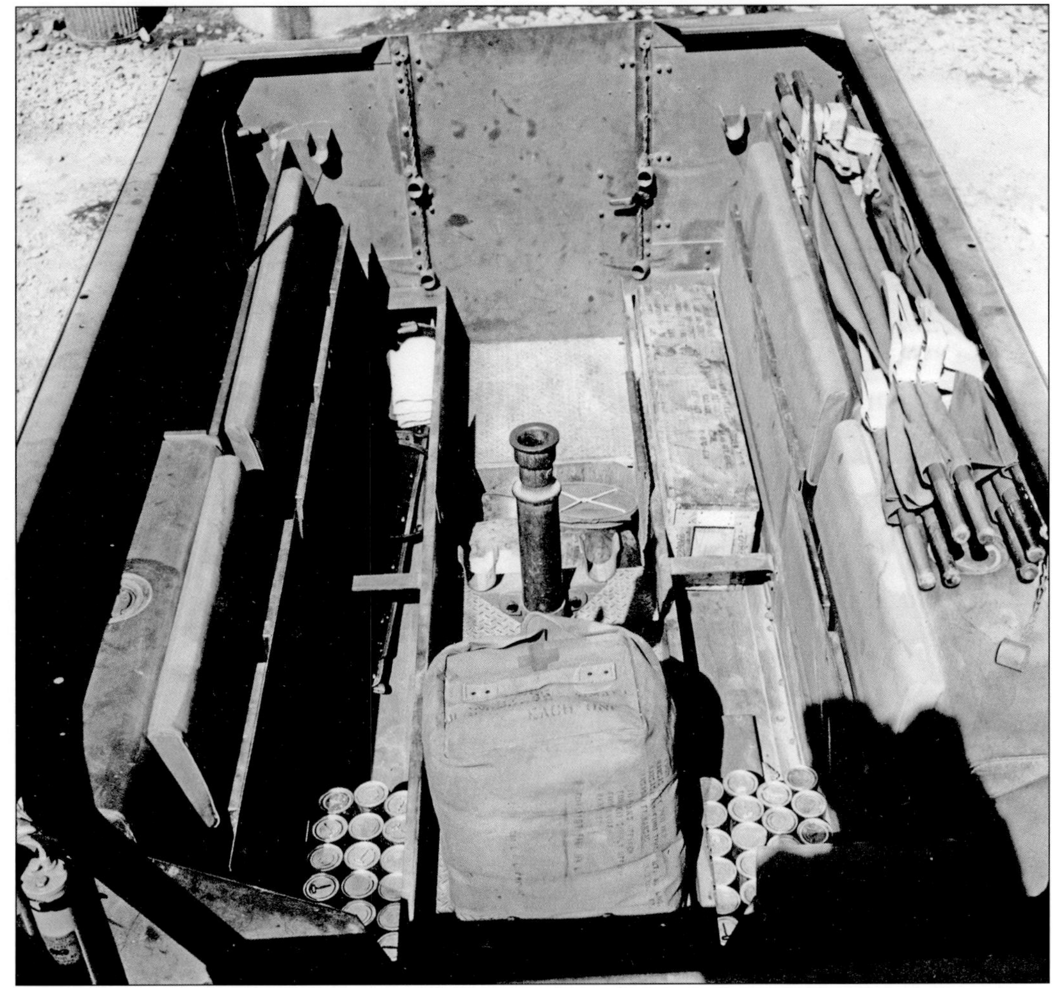

Above: *Medics are lined up in front of two M3 half-tracks used as armored ambulances at Fort Sill, Oklahoma. On the door of the nearer half-track is painted the nickname "MATERNITY." Litters are stowed inside the vehicles. (NARA)*

Below: *In Project 136-8 in early 1943, the Armored Force Board at Fort Knox, Kentucky, tested the stowage provisions for a 4-Litter Carrier (Evacuating), Half-Track, M3. This view to the rear shows the equipment stowed; to the right are the litters. (NARA)*

The 4-Litter Carrier (Evacuating), Half-track, M3 as tested at Ft. Knox is seen from the rear with equipment stowed. The rifle racks had been removed from the rear of the vehicle, and the seat-back frames had been straightened to permit more storage space. (NARA)

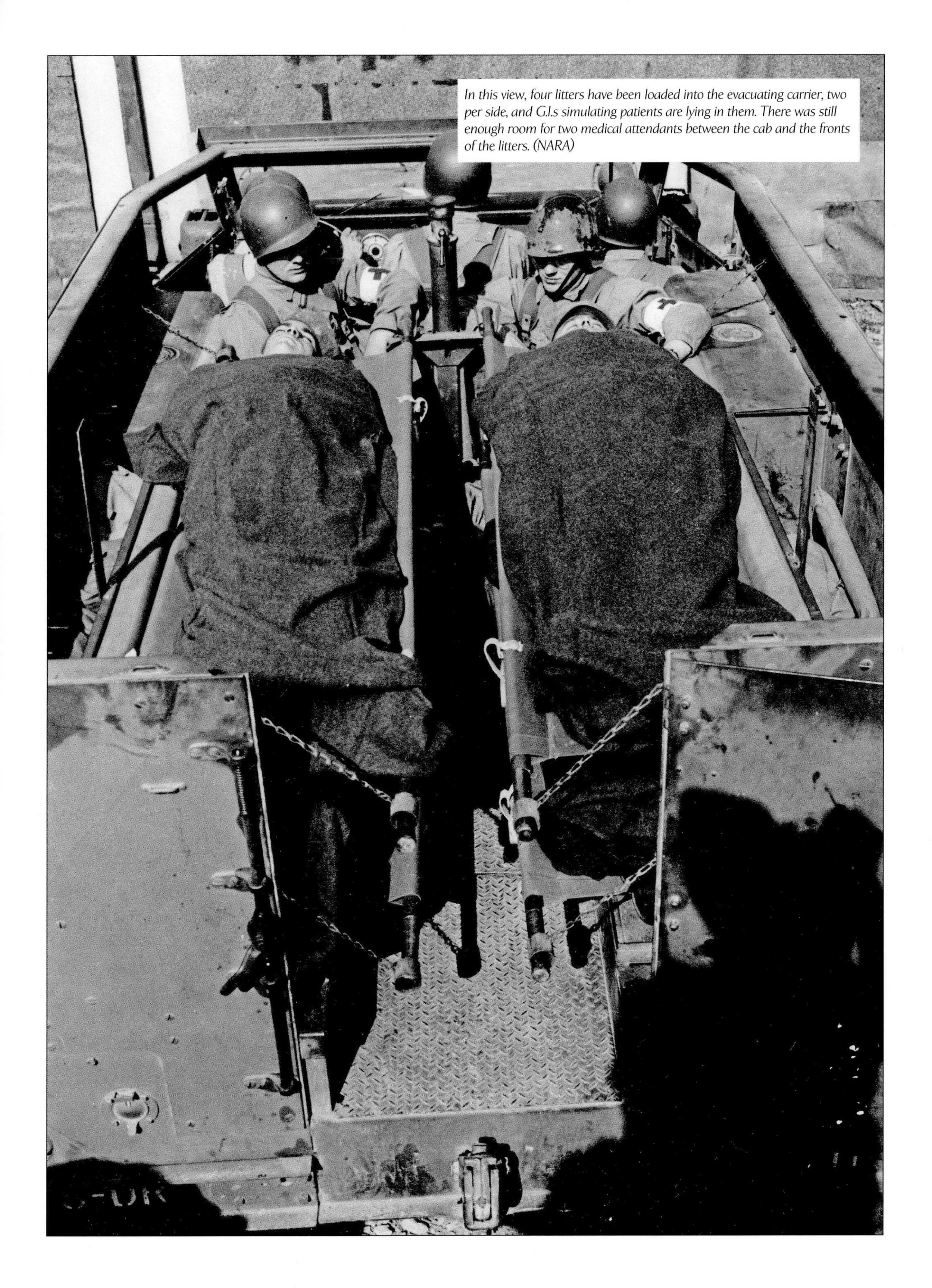

In this view, four litters have been loaded into the evacuating carrier, two per side, and G.I.s simulating patients are lying in them. There was still enough room for two medical attendants between the cab and the fronts of the litters. (NARA)

Above: *In Project 233 in early 1942, the Armored Force Board at Fort Knox, Kentucky, developed attachments for carrying four litters in an M3 half-track, including sockets on chains attached to the body frame, and racks along the centerline.*

Below: *In the summer of 1944 the Armored Force Board tested a four-litter carrying arrangement on an M3A2 half-track. This photo was taken to illustrate a simple, folding step that was added to facilitate entry to the vehicle, although the step is difficult to discern. (NARA, both)*

Above: *During tests designated Project 491 in the summer of 1944, spring-loaded litter brackets were analyzed. As seen here, as an expedient the top litters could be tilted downward and secured in place in order to form seat backs when so required.* (NARA) **Below:** *Medical attendants load a litter patient into an M3 half-track ambulance during an exercise. The flap in the rear curtain of the tarpaulin was especially handy for personnel when handling heavy loads such as this one.* (LOC)

Above: *On this M3 half-track ambulance nicknamed "MID-FORCEPS," the litters are stowed on the outside of the vehicle by means of straps. A large red cross on a white background is painted on the cab door.* **Below:** *Medical attendants of the Fifth Army carefully load a wounded soldier into an M3 half-track ambulance in the outskirts of Rome on 4 June 1944. Three five-gallon liquid containers are stowed on a rack on the rear of the body. (NARA, both)*

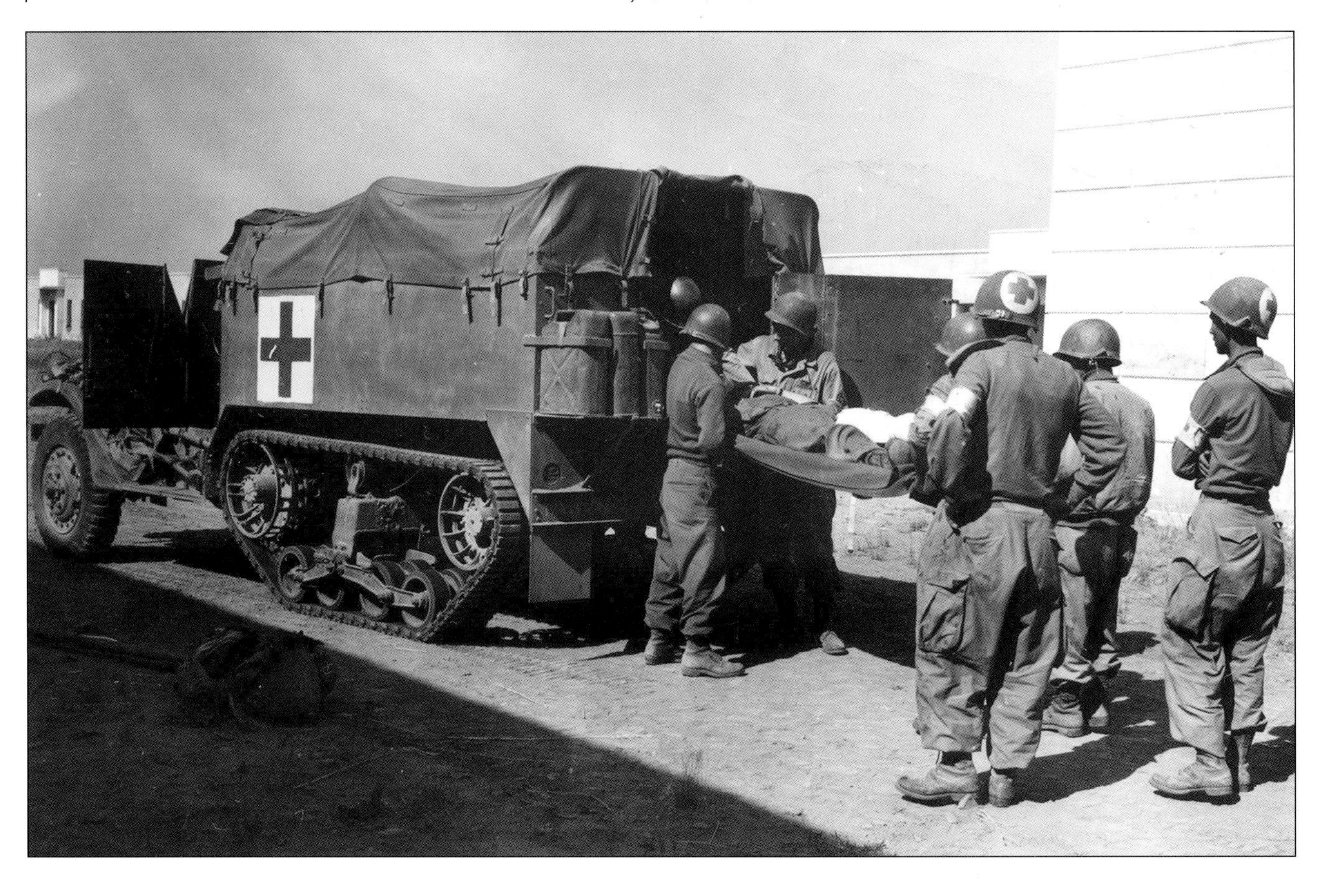

Above: *A wounded G.I. is being placed in an M3 half-track ambulance in the western part of Germany in mid-December 1944. A half-track ambulance had an advantage in mobility in muddy conditions such as these. (NARA)*

Below: *Chains are installed on the tires of this M3 half-track ambulance in the European Theater of Operations during the final winter of World War II. Much equipment is lashed to the front end of the vehicle. (NARA)*

This photograph and the ones on the following page document the installation of a spring-loaded litter rack in an M5A1 half-track ambulance, as tested during Project 431 by the Armored Force Board at Fort Knox, Kentucky, during World War II. (NARA)

Above: *Four empty litters are installed in the M5A1 half-track ambulance. Sockets attached to chains are used to hold the rears of the rails of the litters. Baggage is secured to racks on the rear of the body by webbing straps. (NARA)*

Below: *The same M5A1 half-track ambulance is now loaded with loads simulating litter patients. Aisle space between the litters was virtually nil, with a maximum of 8 3/8 inches available between the two upper litters. (NARA)*

Above: *Project 241 of the Tank Destroyer Board at Camp Hood, Texas, in July 1944 was an experiment to store 28 T15 containers for 3-inch ammunition on the sides of an M3 half-track. Fourteen of the T15 containers were stowed on each side of the body. (NARA)*

Below: *Also during the Project 241 tests, experiments were conducted with storing five T15 ammunition containers in the mine rack on each side of an M3 half-track. Wooden spacers were placed on the rack to keep the containers from shifting during travel. (NARA)*

Above: *In this photo taken at Camp Hood, Texas, during Project 241, the T15 3-inch ammunition containers are stored obliquely. The tops of the containers are secured by a retainer bracket, and the bottoms of the containers are seated on angled blocks. (NARA)*

Right: *Project 241 also tested this method of stowing four T15 containers on the inside of the rear door. A block was fitted under one of the containers to lift it so that the ring on the container didn't touch the rings on the adjacent containers, making for more room. (NARA)*

Above: *Produced in extremely limited numbers, the T17 and T17E1 half-tracks were M3 chassis fitted with the wooden cargo body designed for the CCKW-352 2 ½-ton truck. Rather than an armored cab and hood, sheet metal truck components were used instead.* **Below:** *The result was designated Carrier, Radio, Half Truck, T17 by OCM item 18638. The vehicles were intended specifically to transport the HO-17 shelter containing the SCR-299 radio set and its associated equipment. (NARA, both)*

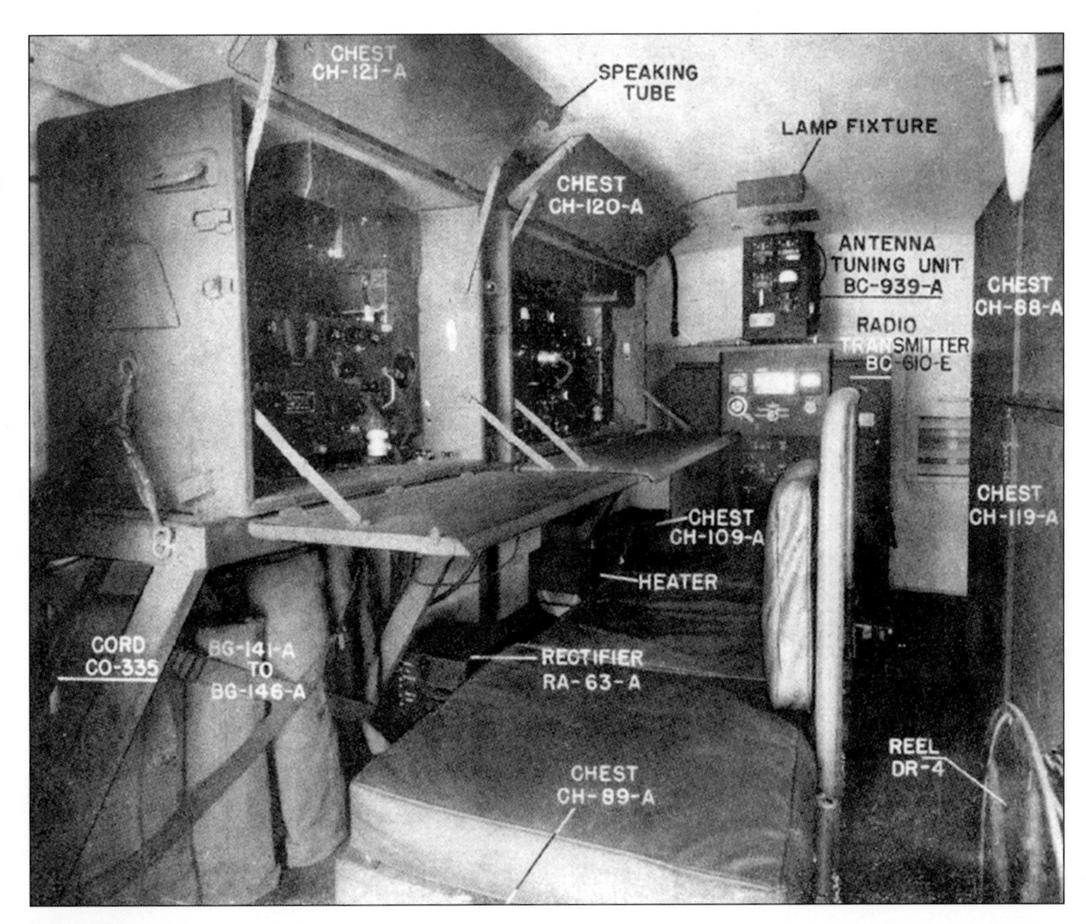

Right: *This technical manual illustration provides a glimpse inside the HO-17 shelter. Due to the presence of so much electronic equipment, the shelter required its own power source. This was typically provided by a modified Ben-Hur trailer known as the K-52 that mounted the PE-95 Power Unit—a gasoline powered generator. However, the substantial overhang of the shelter in this case precluded its use. (NARA)*

Below: *The T17 was not the only "soft-skin" derivative of the G-102 series halftrack. Additionally, three 500 gallon capacity fuel servicing trucks were assembled, registration numbers 4080901 through 4080903. Known alternately as the F-4 or L34, these vehicles, like the T17, utilized Autocar Series 80 cab sheet metal in lieu of armor. (National Museum of the United States Air Force)*

Chapter 10 Post-WWII

The Half-Track Rolls Into the Sunset

The U.S. military began a shift toward full-tracked, fully enclosed armored personnel carriers during WWII. The 1944 Report of the Armored Equipment Board made note of the "...unsatisfactory record of the half track to date..."

It is not surprising then that after VJ-Day many war-weary half-tracks were scrapped in theater. Better machines were retained by occupation troops, or returned home for depot overhaul and continued use as the military's planned transition to full-tracked carriers was implemented.

Two things interfered with this plan. First, the development of affordable, reliable full-tracked carriers took longer than planned. Second, on 25 June 1950, the Communists crossed the 38th parallel in Korea. With war ongoing in Korea, and a third World War seemingly imminent, development of the new generation of personnel carriers was expedited, but at the same time, new life was given to the aging half-tracks, as they were immediately available.

By the late 1950s, as seen earlier in this volume, half-track cars and personnel carriers had been declared obsolete by the U.S. Army.

As seen earlier, the French had been supplied a number of U.S.-produced half-tracks during WWII. The postwar Mutual Defense Assistance Program bolstered those numbers, as well as supplying additional vehicles to other allied nations. Some of the French vehicles would see service in Southeast Asia.

The Commonwealth, who also had received half-tracks during WWII, acquired even more postwar. Postwar, the vehicles were renumbered into the series 45YS13-51YS13 (White), 34YT61-40YT70 (International), and 52YZ46-67YZ19 (International) in 1949.

After the war, REME (Royal Electrical & Mechanical Engineers) notably converted a number to light aid detachment (LAD) fitters' vehicles, equipping them with enlarged superstructures and folding cranes, and using them for replacing AFV power packs. Other roles for which the vehicles were used or converted include:

- Armored personnel carrier
- Tank destroyer-armed with 6-pounder (57mm) gun
- Self-propelled 75mm gun
- Artillery tractor (for 17-pounder gun on wheeled carriage)
- Artillery tractor (for 75mm gun on wheeled carriage)
- Regimental command vehicle
- Radio vehicle
- 3in mortar carrier
- Vickers machine gun carrier
- Recovery vehicle

However, the most notable use of the halftrack following WWII was in the hands of the Israelis. The newly formed Jewish state began acquiring the vehicles in 1947, with the initial purchases being made

After the end of the war in Europe, U.S. Army vehicles by the thousands were prepared for shipment to the Pacific in preparation for the planned invasion of Japan. Here, Cpl. Fred W. Guyette and Sgt. Andrew McGann renew the markings on a rare M3A1 Scout Car being processed at Marseilles, France, for shipment to the Pacific on July 19, 1945. "Prestone 45" on the radiator armor indicates that antifreeze had been added to the radiator in 1945. (NARA)

PRESTONE 45
USAW 6089815-S

in Great Britain of International Harvester-produced vehicles. The overflowing yards of surplus dealers in the UK would continue to be a source for Israeli half-tracks (and M3A1 scout cars) into the 1950s.

In Israeli service the vehicles were subjected to extensive modification. Rear doors were added to M2 series vehicles, a variety of armament upgrades and modifications were employed, and the power-plants subjected to considerable change. In some instances, International Harvester Red 450-B engines were installed in half-tracks produced by Autocar/Diamond T/White. In the late 1970s these engines began to be replaced with Detroit Diesel 6V53 Diesel engines. This provided a parts commonalty with the M113A1 personnel carriers making their way in increasing numbers in the IDF, had a greater parts availability than the WWII-era power plants, burned a less volatile fuel, and had increased torque over the aging gasoline engines. This conversion often included in installation of rear-mounted external fuel tanks, providing more space inside the vehicle, as well as moving the fuel outside of the crew compartment.

The long service life of the U.S.-made half-track in Israeli service, stretching into the 21st century, has been a boon to the enthusiast. Due to this military service, fresh stocks of the somewhat fragile tracks continued to be produced in to the 1970s.

Above: *Men are cutting armor plate from the front end of a U.S. half-track, registration number 405007, at a base at Etain, France, on 2 February 1946. It was one of many vehicles being repaired at this base prior to being sold as war surplus. (NARA)*

Below: *During maneuvers at Fort Knox, Kentucky, members of an armored infantry unit in a half-track take a break before heading out on a reconnaissance mission. In the mine racks are six M6A2 heavy antitank mines. (NARA)*

U.S.A.
405007 N
0841319

Above: *This vehicle has had a varied life. According to its registration number, it was built as a M3. Later it was updated to M3A1 standards through installation of the ring mount and pulpit. By the time this photo was taken in the early 1950s at Grafenwoehr, the ladder-like stowage brackets and tarp hoops engineered for the aborted M3A2 series were installed. (Bob Wiebold via John Adams-Graf)* **Below:** *A winch-equipped M3A1 half-track rolls down the ramp of LST-397 at a landing beach on Vieques Island, Puerto Rico, during Operation Portrex, a two-week-long exercise during March 1950 to evaluate joint-service doctrine. (US Army Engineer School History Office)*

Above: *A front roller-equipped M3A1 patrols the roads of South Korea shortly before the outbreak of hostilities in June 1950. Typical of garrison troops, the men are outfitted very lightly, with some sporting ascots.* **Below:** *The crew of an M3A1 half-track of the 15th Antiaircraft Artillery Automatic Weapons Battalion, 7th Division, fire on an enemy position on the outskirts of Yanggu, Republic of Korea, on 15 April 1951. The nickname "Macky" is painted on the body. (NARA, both)*

Above: *A Sergeant Veale of the Royal Canadian Signals leans on the driver's door of an M3A1 wireless half-track assigned to Tactical Headquarters in Korea on 7 November 1951. A piece of electrical equipment with a cover over it is on a rack on the fender. (NARA)* **Below:** *One of the more unusual uses of a U.S. half-track in foreign use is this machine in use in Finland in the 1950s. In order to make maximum use of the vehicle's mechanical abilities, the rear armor has been removed and replaced with a simple wooden stake body. Completing the unusual configuration is its load; a German sfH 18 15cm artillery piece. (SA-kuva-arkisto)*

Above: *The Israelis acquired half-tracks from a number of sources, both just before and during the War of Independence in 1948. By far, the vast majority of these were IHC versions, like these M5s. There were also many foreign volunteers in the IDF, including this 18 year-old American, Ralph Lowenstein. He was a driver in the 79th Armored Battalion, which fought on the northern front. (Ralph Lowenstein, via Tom Gannon)*

Below: *The lettering on the side indicates that this vehicle's original postwar destination was Letterkenny Army Depot in Pennsylvania, to be repatriated following Canadian service. It was obviously diverted. The soldier sitting on the side is Hertz Hirshberg, the young Israeli commander of a platoon of modified M5s converted to carry the QF (Quick-Firing) 6-Pounder anti-tank gun. (Hertz Hirshberg, via Tom Gannon)*

Above: *Half-tracks were the primary APC for the IDF through the '50s and '60s. This is a group of officer candidates going through training in 1966. The IDF used M2s, M2A1s, M3s, M3A1s, M5s, M5A1s and M9A1s in a variety of configurations, beyond the basic M3 personnel carrier version shown here. The .30-caliber machine gun is mounted in a standard Sherman ball mount. (Eran Kaufman, via Tom Gannon)* **Below:** *This is a modified M2 manned by Israeli reservists in Lebanon in 1982. To make more room in this short-hull version of the series, the fuel tanks were moved to the rear, using an indigenous design. Also, the old gasoline-powered engine was replaced by the diesel engine from the M113 series. Internal modifications and additional armor protection on the front were also features of the upgrade. (Frans van Stalkenburg, via Tom Gannon)*

Left: *As testament to the reliability and usefulness of the design, this White/ Diamond T half-track was still in use into the '90s at least. It mounts a Tzefa mine-clearing rocket. Published photographs show this type of half-track in use alongside the latest version of Israeli-modified Patton series, the MAGACH 7, and the Merkava. (Public display, via Tom Gannon)*

Below: *A few French half-tracks found their way to Vietnam after World War II. By the time of U.S. involvement in Vietnam in the 1960s and early 1970s, some of these half-tracks were still in use by ARVN forces. This M3, photographed in February 1969, has an extra headlight on the left side and numerous stowage brackets on the side of the body. (James L. Loop)*

Above: *Stock features of an M3 visible on the rear of this example include the taillight assemblies, receptacle for trailer lights and brakes connection, door latch assembly, and grab handles. Early-type, fender-mounted headlights and brush guards are also present.* **Below:** *This dilapidated M2 or derivative was performing static defense duties when photographed in the Republic of Vietnam in early 1969. A .30-caliber machine gun with a small shield and a .50-caliber machine gun are under wraps. The winch has been removed, and most of the radiator shutters are missing. Note the spare tire behind the passenger's seat. (James L. Loop, both)*

Above: *Among other nations, Mexico used American-built half-tracks long after WWII. This particular military parade contained M2A1s and M3A1s. Especially interesting is the modified front end. The Mexican government contracted with an Israeli contractor to initiate engine improvements, similar to those used in the IDF upgraded versions. Included was the Diesel engine, along with the headlight assembly from the M113. (Mario Martinez, via Tom Gannon)* **Below:** *Chile acquired U.S. half-tracks postwar as well. Aside from a few local modifications, such as the spare tire rack, the Chilean army independently acquired the Eyal crane assembly used on Israeli fitter-type half-tracks, M113s and various trucks, which were installed in army depots. (Fernando Wilson, via Tom Gannon)*

Appendix A

Production, Registration and General Data

M3A1 Scout Car

Qty	Year	Contract	PO Number	Lot#	Registration number range	Ord. serial number range
24	1939	W-741-ORD-3905	—	1	—	106-111, 113, 115, 117, 133-147
274	1939	—	X.O. 5520	1	—	— **(A)**[1]
25	1940	W-741-ORD-7128	PA-2503-2595	—	—	— **(B)**
256	1940	W-741-ORD-4345	PA-4003	—	—	—
13	1940	W-741-ORD-5296	T-117A	1	—	405 - 417 **(C)**[2]
3	1940	W-741-ORD-5680	T-175	2	—	418 - 420
1,054	1940	W-741-ORD-5680	T-278A	2	—	421 - 1474 **(D)**
1,075	1941	W-741-ORD-6099	T-422	2	—	1475 - 2549 **(E)**
282	1941	W-741-ORD-6258	T-527	2	—	2550 - 2831 **(F)**
32	1941	W-741-ORD-5680	T-915	2	—	2831 - 2863 **(G)**
107	1941	W-303-ORD-1042 [1]	T-2112	3A	609819-609925	2865 - 2971 **(H)**
600	1941	DAW-303-ORD-86	T-2382	3	609219-609818	2972 - 3571 **(I)**
5,005	1941	W-303-ORD-1181	T-3146	5	6011067-6016071	3572 - 8576 **(J)**
800	1942	DAW-303-ORD-4026	T-3160	4	—	8577 - 9376 **(K)**
26	1942	W-303-ORD-1400	T-3599	5	604690-604715	19377 - 19402
10,000	1942	W-303-ORD-1260	T-3338	6/7[3]	6082640-6092639	9377 - 19376 **(L)**
1,623	1942	W-303-ORD-2077	T-4297	8	6072773-6074395	19403 - 21025

A. Completed in 1939, 100 to be Diesel powered per OCM 15757
B. Completed April 1940
C. Completed in July 1940
D. 145 Oct. 40; 162 Nov. 40; 231 Dec. 40; Jan. 41, 376 Jan. 41, 143 Feb. 41
E. 232 Feb. 41; 414 March 41; 326 April 41; 26 Oct. 41; 61 Nov. 41; 13 Dec. 41
F. 98 April 41; 184 May 41
G. All built March 1941
H. 100 May 41; 7 June
I. 243 March 42; 357 April 42
J. 293 April 42; 800 May; 1187 June; 1,200 July; 1,200 Aug.; 325 Sept.
K. 800 May 42
L. 75 May 42; 1,200 June; 1200 July; 1,200 Aug.; 1,200 Sept.

1. Data per Tank Automotive Production Report December 1941
2. ATCV erroneously lists as W-303-ORD-1074
3. 4,000 of these vehicles were lot 6, 6,000 were lot 7

Contract value and duration:

Contract W-741-ORD-5680	$5,621,000	period 6-40 through 3-41
Contract W-741-ORD-6099	$5,550,000	period 8-40 through 12-41
No contract number	$260,000.00	period 8-40 through 12-41
Contract W-741-ORD-6258	$1,488,000.00	period 10-40 through 5-41
Contract W-741-ORD-7128	$284,000.00	period 12-40 through 10-41
Contract W-303-ORD-1042	$569,000.00	period 9-41 through 4-42
Contract DAW-303-ORD-86	$3,221,000.00	period 10-41 through 9-42
Contract W-303-ORD-1181	$32,122,000.00	period 1-42 through 3-43
Contract W-303-ORD-1260	$52,720,000.00	period 1-42 through 2-44
Contract W-303-ORD-181L	$5,159,000.00	period 1-42 through 12-42
Contract W-303-ORD-1400	$170,000.00	period 4-42 through 8-42
Contract W-303-ORD-2077	$9,281,000.00	period 9-42 through 3-44
Contract W-741-ORD-10612	$153,000.00	period 10-41 through 2-42
No contract number	$105,000.00	period 10-41 through 5-42
Contract W-741-ORD-11267	$73,000.00	period 12-41 through 6-42
Contract W-303-ORD-1280	$65,000.00	period 1-44 through 7-44

M2/M2A1 Half-track Car

Qty.	Year	Contract	PO Number	Model/Make	Registration number range	Ord. serial number range
537	1941	W-741-ORD-6276	T-515	M2/Autocar	—	2 - 538
4,629	1941	W-741-ORD-6285	T-540	M2/White	4010200 - 4014829	539 - 5167
347	1941	W-303-ORD-977	T-1685	M2/White	4016710 - 4017056	5447 - 5793
1	1941	W-670-ORD-1989	T-3292	M2/Autocar	4019531	5794
806	1941	W-670-ORD-1989	T-3292	M2/Autocar	4018413 - 4019218	5795 - 6600
1,400	1941	W-670-ORD-1989	T-3339	M2/Autocar	4019532 - 4020931	6601 - 8000 [1]
2,000	1942	W-303-ORD-1313	T-3511	M2/White	4022934 - 4024933	8001 - 10000 [1]
239	1941	W-670-ORD-1989	T-3512	M2/Autocar	4024934 - 4025849	10001 - 10916 [1]
677	1941	W-670-ORD-1989	T-3512	M2A1/Autocar*	4024934 - 4025849	10001 - 10916 [1]
250	1942	W-303-ORD-1611	T-3845	M2/White	4036766 - 4037015	12512 - 12761 [3]
250	1942	W-303-ORD-1611	T-3846	M2/White	4037016 - 4037265	12762 - 13011 [4]
89	1942	W-303-ORD-4973	T-4921	M2A1/White	40100994 - 40101082	16212 - 16300 [1]
20	1942	W-303-ORD-4972	T-5243	M2A1/White	40103434 - 40103453	16301 - 16320 [1]
9	1943	W-670-ORD-4431	T-7115	M2/Autocar	40111257 - 40111265	16321 - 16329 [2]
947	1942	W-303-ORD-2080	T-4333	M2/White	4078165 - 4079968	17986 - 19789 [1]
857	1942	W-303-ORD-2080	T-4333	M2A1/White*	4078165 - 4079968	17986 - 19789 [1]
2	1942	W-271-ORD	T-3762	M2E5/IHC	4015156 - 4015157	4958 - 4959

1. 50% with winch, 50% with roller
2. Without armament, special for Navy
3. With front-mounted roller
4. With front-mounted self-recovery winch
*Included in M2 listing for same PO above

Autocar contracts value and duration:

Contract W-741-ORD-6276	$8,959,000.00	period 9-40 through 4-42
Contract W-670-ORD-1989	$27,262,000.00	period 2-42 through 2-45
Contract W-670-ORD-4431	$98,000.00	period 5-43 through 8-43

White contracts value and duration:

Contract W-741-ORD-6285	$36,052,000.00	period 10-40 through 9-42
Contract W-303-ORD-977	$2,256,000.00	period 6-41 through 4-42
Contract W-303-ORD-1313	$12,698,000.00	period 2-42 through 2-43
Contract W-303-ORD-1611	$5,924,000.00	period 5-42 through 6-43
Contract W-303-ORD-4972	Data unknown	—
Contract W-303-ORD-4973	Data unknown	—
Contract W-303-ORD-2080	$13,159,000.00	period 9-42 through 3-44

M3 Personnel Carrier

Qty	Year	Contract	PO Number	Make	Registration number range	Ord. serial number range
520	1941	W-741-ORD-6290	T-515	Autocar	400400 - 400919	2 - 521
641	1941	W-741-ORD-6290	T-544	Autocar	401141 - 401781	1263 - 1903
312	1942	W-670-ORD-1989	T-3257	Autocar	4019219 - 4019530	6160 - 6471
203	1942	W-670-ORD-1989	T-3683	Autocar	4026935 - 4027137	10072 - 10274 [1]
4	1943	—	T-4257	Autocar	—	34930 - 34933 [2]
2,000	1941	W-741-ORD-6289	T-545	Diamond T	401782 - 403781	1904 - 3903
610	1941	W-271-ORD-600	T-1683	Diamond T	4027649 - 4028258	4304 - 4913 [3]
600	1942	W-271-ORD-965	T-3349	Diamond T	4045909 - 4046508	6472 - 7071
1,000	1942	W-271-ORD-964	T-3340	Diamond T	4043329 - 4044328	7072 - 8071 [4]
238	1942	W-271-ORD-1023	T-3513	Diamond T	40111266 - 40111503	8072 - 8309 [5]
800	1942	W-271-ORD-1023	T-3513	Diamond T	40102634 - 40103453	8310 - 9109 [6]
982	1942	W-271-ORD-1385	T-3720	Diamond T	4028260 - 4029241	10275 - 11256 [7]
1,000	1942	W-271-ORD-1967	T-4140	Diamond T	4054460 - 4055459	22831 - 23830 [8]
179	1942	W-271-ORD-2642	T-4571	Diamond T	40104254 - 40104432	30121 - 30299 [9]
1	1943	W-271-ORD-853	T-3133	Diamond T	—	—
100	1941	W-303-ORD-945	T-1295	White	409482 - 409581	3904 - 4003
12	1942	W-303-ORD-945	T-3982	White	—	—
161	1942	W-303-ORD-1611	T-3844	White	4037266 - 4037426	18778 - 18938

1. With winch
2. With cargo bodies
3. 595 converted to M3A1
4. Converted to M3A1
5. 179 w/winch, 59 w/roller
6. 600 w/winch, 200 w/roller
7. 560 w/winch, 422 w/roller
8. 750 w/winch, 250 w/roller
9. 135 w/winch, 44 w/roller

M3A1 Personnel Carrier

Qty	Year	Contract	PO Number	Make	Registration number range	Ord. serial number range
1,000	1942	W-670-ORD-2842	T-4103	Autocar	40109110 - 40110109	21831 - 22830 [1]
1,289	1942	W-670-ORD-3003	T-4661	Autocar	4065202 - 4066490	26941 - 28229 [2]
113	1944	36-034-1417	T-11019	Autocar	Retained original number	40128 - 40240 [3]
1,247	1944	36-034-1878	T-13054	Autocar	Retained original number	47636 - 48882 [4]
3,110	1942	W-271-ORD-2432	T-4659	Diamond T	4062092 - 4065201	23831 - 26940 [5]
273	1942	W-271-ORD-2727	T-4922	Diamond T	40104433 - 40104705	30300 - 30572 [6]
30	1942	W-271-ORD-2726	T-6971	Diamond T	40111227 - 40111256	30848 - 30877 [7]
281	1944	—	T-11453	Diamond T	Retained original number	40241 - 40521 [8]
595	1944	11-022-3630	T-13146	Diamond T	Retained original number	Retained original number [9]
1,045	1944	04-200-202	T-13146	Diamond T	Retained original number	Retained original number [10]

1. 800 w/winch, 200 w/roller
2. 967 w/winch, 322 w/roller
3. Remanufactured from M3 GMC
4. Remanufactured from M3 GMC
5. 2,333 w/winch, 777 w/roller, 8 special
6. 205 w/winch, 68 w/roller
7. With winch, without armament
8. Remanufactured by Chester Tank Depot from T48
9. Converted from M3 by Wylie, Inc.
10. Converted from M3 by Bowen and McLaughlin

The reader should note that the above information on the M3 and M3A1 is based on data appearing in the 1944 *Ordnance Department Armored, Tank and Combat, Vehicles, 1940, 1941, 1942, 1943, 1944*. However, detailed examination of many surviving vehicles indicates it is largely erroneous in regard to the assignment of registration and serial numbers.

M3E2/M5 and M5A1 Personnel Carrier

Qty.	Year	Contract	PO Number	Model/Make	Registration number range	Ord. serial number range
1	1943	W-271-ORD-1452	T-3762	M3E2	403307	3429
1	1943	W-271-ORD-1452	T-3762	M3E2	409422	3844
7,519	1943	W-271-ORD-1426	T-3781	M5	4029247 - 4036765	1259 - 18777
65	1943	W-271-ORD-2769	T4936	M5A1	40101631 - 40101695	8070 - 8134

M2E5 and M9A1 Car, half-track

Qty.	Year	Contract	PO Number	Model/Make	Registration number range	Ord. serial number range
2	1943	W-271-ORD-1452	T-3762	M2E5/IH	4015156 - 4015157	4958-4959
3,433	1942	W-271-ORD-2643	T-4660	M9A1/IH	4067586 - 4071018	3- 3435*

* 4,286 w/winch, 3,233 w/roller per IH data (Ordnance has 4,287 w/winch; 3,232 w/roller).

** 37 w/winch, 28 w/roller per IH data (Ordnance has 36 winch/29 roller). This order, issued on 7 November 1942, originally called for 1,000 M5 equipped as ambulances, and on 16 December 1942 was modified to specify standard M5 vehicles. On 29 December 1943 the order quantity was reduced to 65 vehicles.

International Harvester M5 and M5A1 contracts value and duration:

Contract W-271-ORD-1426	$61,666,000.00	period 4-42 through 3-44
Contract W-271-ORD-2769	$8,366,000.00	period 11-42 through 1-44*
Contract W-271-ORD-2770	$4,039,000.00	period 11-42 through 9-43

* On 2 November 1943 the Chicago Ordnance District issued supplement number three to contract W-271-ORD-2769/2770, which formally combined the two letter purchase orders for administrative purposes, and at the same time reduced the quantify of vehicles on this order from 1,548 to 1,045, valued at $10,251,750.42. On 29 December 1943 the Chicago Ordnance District issued a partial notice of termination, further reducing the quantity of vehicles to 65 (28 M5A1 with roller and 37 M5A1 with winch), with a similar reduction in spare parts. This reduced the value of the combined 2769/2770 contracts to $600,547.17. Contract originally included 1,000 ambulance conversion kits, in addition to vehicles.

International Harvester M9A1 contracts value and duration:

Contract W-271-ORD-2643	$26,670,000.00	period 10-42 through 12-43

Halftrack General Data

Model	M2	M2A1	M3
SNL Group	G-102	G-102	G-102
Make	White, Autocar	White, Autocar	White, Autocar, Diamond T
Weight, Lbs [1]	19,195	19,600	20,000
Length w/Winch [2]	241.625	241.625	249.625
Length W/Roller [3]	234.75	234.75	242.625
Height	89	100	89
Width, Body	77.25	77.25	77.25
Width, with Mine Racks	87.5	87.5	87.5
Side/Rear Armor	.25	.25	.25
Tire Size	8.25 - 20	8.25 - 20	8.25 - 20
Engine	White 160 AX	White 160 AX	White 160 AX
Fuel	Gasoline	Gasoline	Gasoline
Fuel Capacity, Gallons	60	60	60
Range, Miles	210	210	210
Electrical	12 negative	12 negative	12 negative
Comb. Trans./Transfer	Spicer 3461	Spicer 3461	Spicer 3461
Transmission Speeds	4	4	4
Transfer Speeds	2	2	2
Turning Radius Feet	29.5	29.5	29.5

Model	M3A1	M5	M5A1	M9A1
SNL Group	G-102	G-147	G-147	G-147
Make	White, Autocar, Diamond T	International	International	International
Weight, Lbs [1]	20,500	20,500	21,500	21,200
Length w/Winch [2]	249.625	241.625	241.625	241.625
Length W/Roller [3]	242.625	242.19	242.19	242.19
Height	106	91	108	108
Width, Body	77.25	77.375	77.375	77.375
Width, with Mine Racks	87.5	86.875	86.875	86.875
Side/Rear Armor	.25	.31	.31	.31
Tire Size	8.25 - 20	9.00 - 20	9.00 - 20	9.00 - 20
Engine	White 160 AX	IH Red 450B	IH Red 450B	IH Red 450B
Fuel	Gasoline	Gasoline	Gasoline	Gasoline
Fuel Capacity, Gallons	60	60	60	60
Range, Miles	210	200	200	200
Electrical	12 negative	12 negative	12 negative	12 negative
Comb. Trans./Transfer	Spicer 3461	IH 1856	IH 1856	IH 1856
Transmission Speeds	4	4	4	4
Transfer Speeds	2	2	2	2
Turning Radius Feet	29.5	29.5	29.5	29.5

1. Combat load
2. Winch equipped vehicles are 430 lbs heavier than roller equipped vehicles.
3. All dimensions listed in inches.

Engine Data

SNL Group	G-102	G-147
Engine Make/Model	White 160 AX	IH Red 450b
Number of Cylinders	6	6
Cubic Inch Displacement	386	450
Horsepower	127 @ 3000	130 @ 2600
Torque	325 @ 1200	348 @ 800
Governed Speed (rpm)	Not Governed	Not Governed

During the early 1950s, the army ordered additional replacement half-track engines from White. Here, one of those new-production engines is tested on a dynamometer at Aberdeen Proving Ground (NARA via Dana Bell)

Appendix B

Notable Half-track Military Work Orders and Illustrations, TOE

G-102 half-track MWOs

Number	Date	Action	Reason
G102-W1	10-Nov-41	Provided stronger drag link	Improve service life
G102-W2	16-Feb-42	Discard cab ventilator	—
G102-W3	4-Apr-42	Provide lubrication chart holder	—
G102-W4	28-May-42	Provide air cleaner shrouds	—
G102-W5	16-Jul-42	Convert electric brakes from 12 to 6 volt	—
G102-W6	25-Jul-42	Stabilize radiator bottom armor plate	—
G102-W7	22-Aug-42	Equip with attachments for carrying litters	—
G102-W8	31-Aug-42	Added surge tank	Improve cooling
G102-W14	29-Nov-42	Spring-loaded idler	Reduce track throwing
G-102-W20		Install black out driving lights	Improve black out driving capabilities
G102-W21	24-Feb-43	Added mine racks	Increased combat effectiveness
G102-W22	25-Feb-43	Added hand grips	Prevent injury
G102-W23	17-Mar-43	Added heavier D48250 volute springs	Prevent breakage
G102-W24	21-Apr-43	Cutaway portion of hood armor	To prevent tire scuffing
G102-W26	18-May-43	Incorporated tapered bogie wheel flanges	Reduce track throwing
G102-W29	26-Dec-42	Increase mortar traverse	Improved combat effectiveness for M4A1
G102-W34	21-Jun-43	Added demountable headlights	Reduce muzzle blast damage on GMCs
G102-W35	6-Aug-43	Added heavier front springs	Prevent breakage
G102-W36	15-Jul-43	Install factory-type idler	To provide spring-loaded idler adjustment
G102-W37	15-Jul-43	Use Protectoseal fuel tank cap	Reduce water intrusion
G102-W40	14-Aug-43	Added clearance to idler shackle	To preclude freezing of idler post and shaft
G102-W43	27-May-44	Added stowage brackets MGMC	To provide 3 shovels & 2 mattocks for digging in vehicles.
G102-W44	—	Added guard	To prevent M15A1 guns from firing into cab.

G-147 half-track MWOs

Number	Date	Action	Reason
G147-W2	14 July 1943	Add pressure relief valve	Prevent blowing bogie wheel seals with grease gun
G147-W3	12 Aug-43	Add rear axle outer bearing seal	To prevent gear lube from entering brakes
G147-W4	12 Nov.-43	Install litter carrier brackets	Convert certain M5 vehicles for use as ambulance
G147-W5	14 Aug.-43	Revise idler shackle	To prevent binding
G102-W5	25 Aug. -43	Reinforce louver control lever	To prevent bending

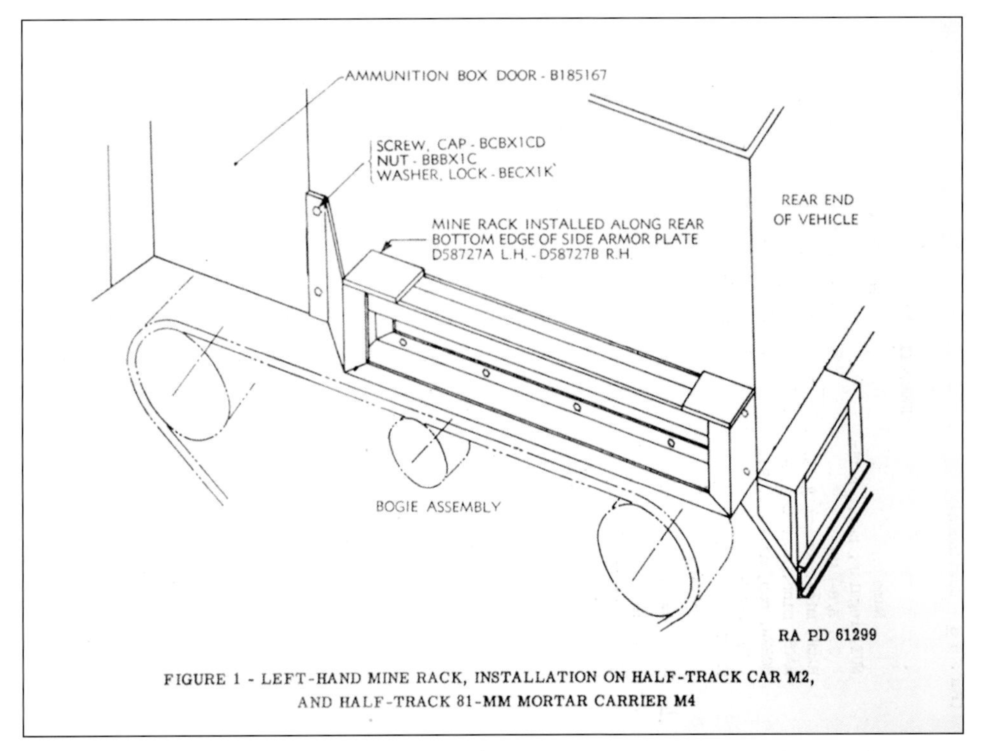

FIGURE 1 - LEFT-HAND MINE RACK, INSTALLATION ON HALF-TRACK CAR M2, AND HALF-TRACK 81-MM MORTAR CARRIER M4

Above: *Illustration from G102-W21, 24-Feb-43.* **Below:** *Illustration from G102-W22, 24-Feb-43.*

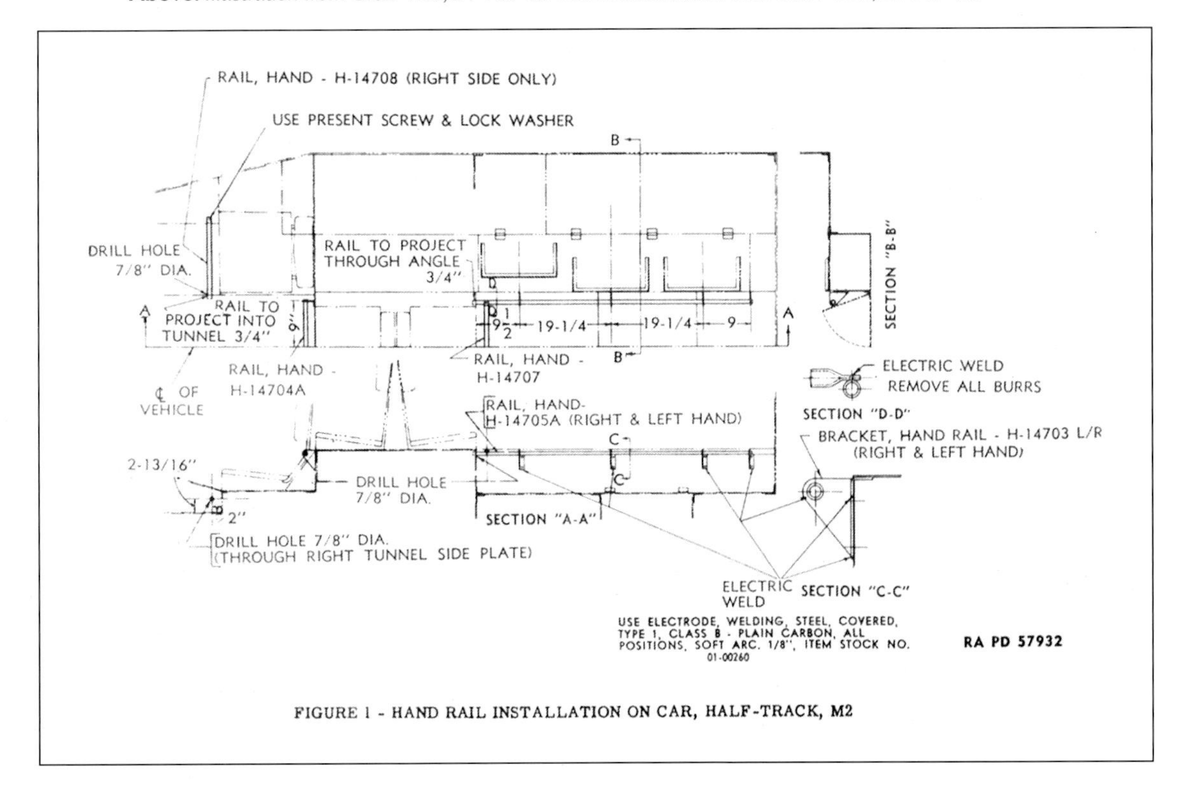

FIGURE 1 - HAND RAIL INSTALLATION ON CAR, HALF-TRACK, M2

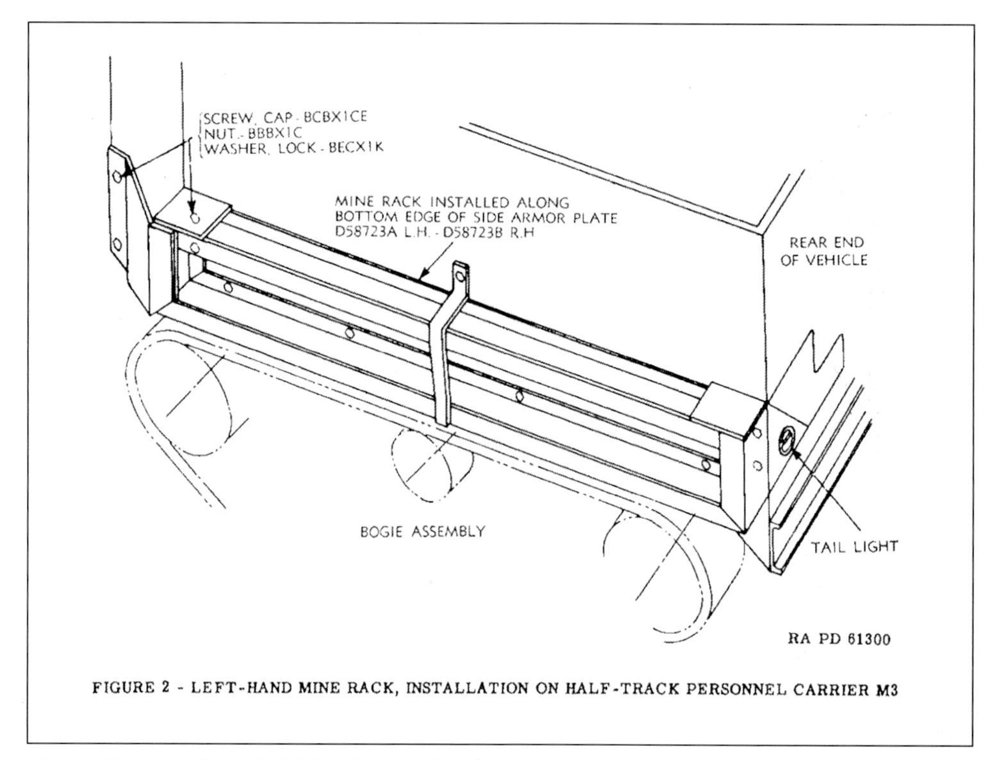

FIGURE 2 - LEFT-HAND MINE RACK, INSTALLATION ON HALF-TRACK PERSONNEL CARRIER M3

Above: *Illustration from G102-W21, 24-Feb-43.* **Below:** *Illustration from G102-W22, 24-Feb-43.*

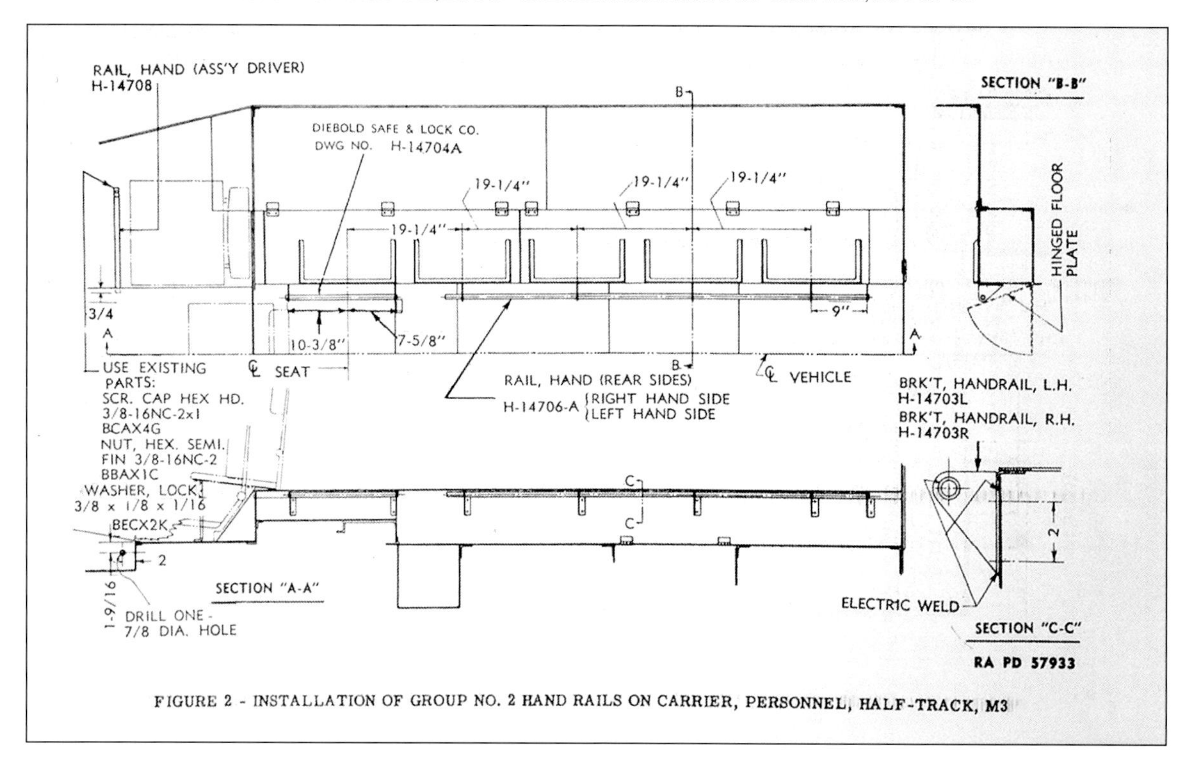

FIGURE 2 - INSTALLATION OF GROUP NO. 2 HAND RAILS ON CARRIER, PERSONNEL, HALF-TRACK, M3

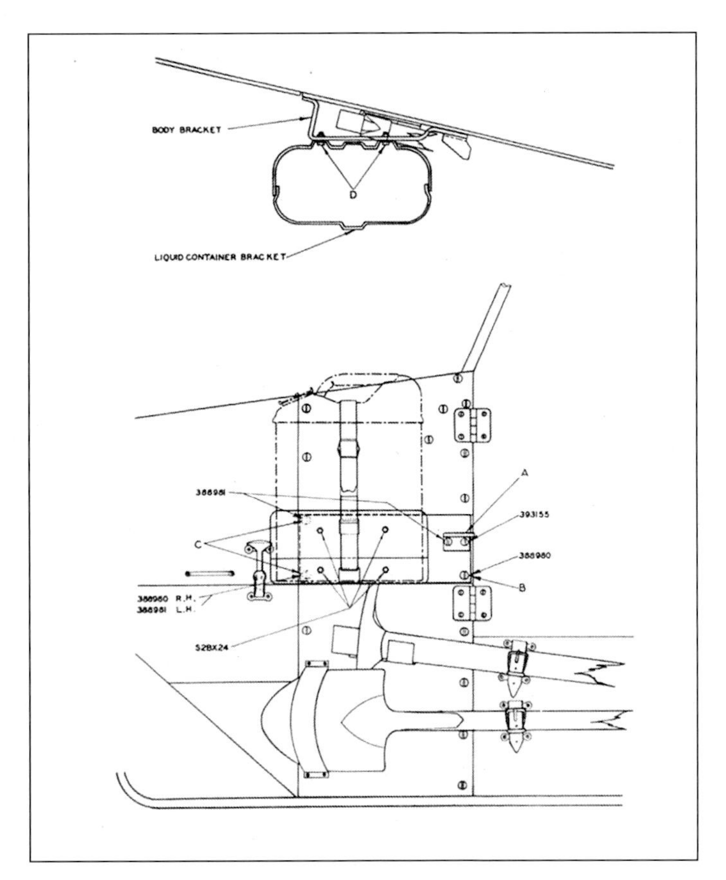

Above: *Illustration from White Service Bulletin number 27, 12 November 1942 specifying the installation of racks for 5-gallon liquid containers.* **Below:** *Illustration from G102-W14, 29-Nov-42.*

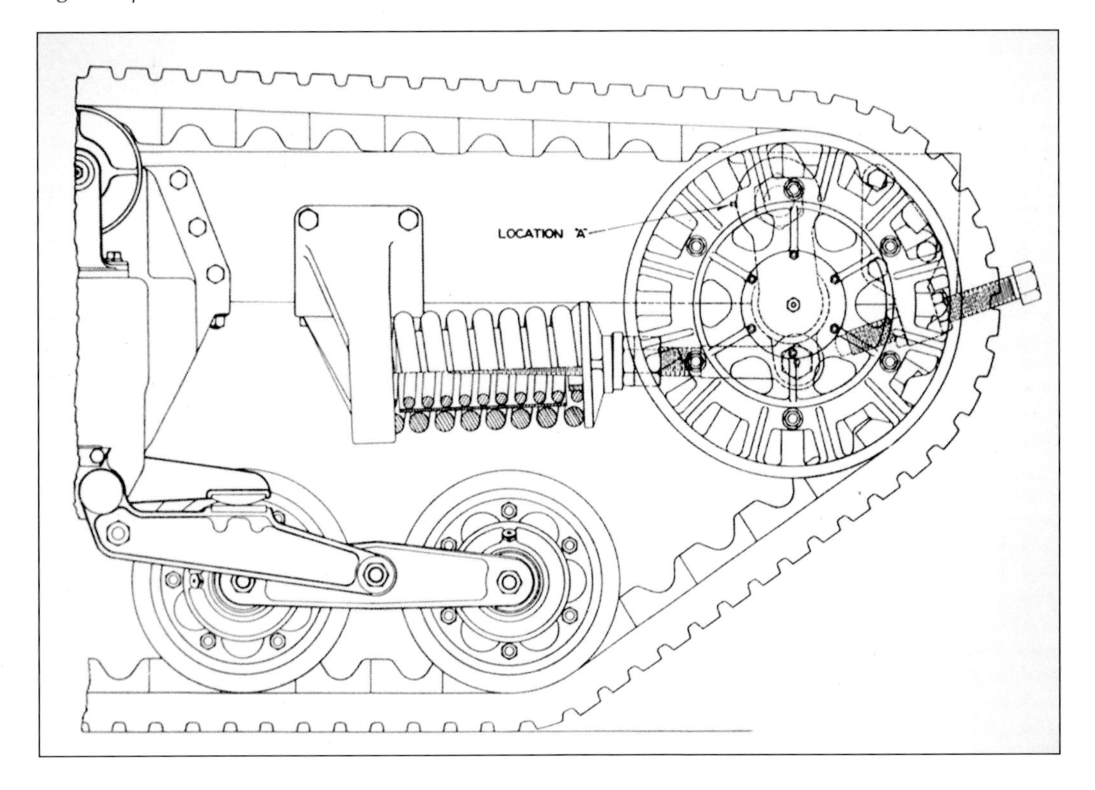

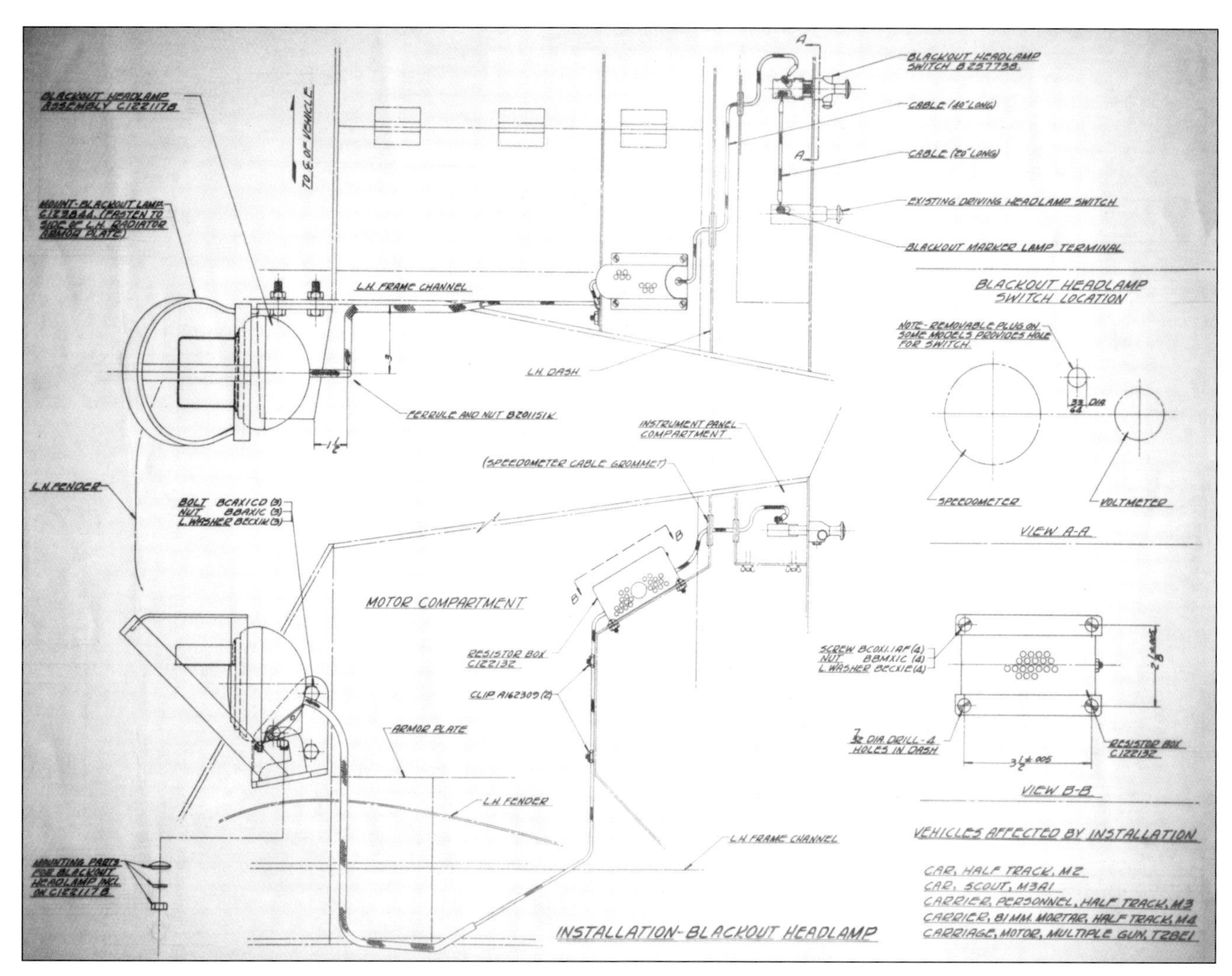

Above and right: *Illustration from MWO G-102-W20 for installation of black out driving lights and photographs from APG test report dated 5 September 1942.*

Half-track TOE data

Infantry division (as 15 July 1943)	5 M3
Infantry Regiment or Battalion, typical (as of 15 September 1943)	78 M3 or M21 mortar carrier
Armored Division (as of 15 September 1943)	
Headquarters Company	16 M3
CCA, CCB, CCR combined	14 M3
Signal Company	19 M3
Cavalry Recon Sq, Mech	32 M3
Tank Battalions	13 M3
Armored Infantry Battalion	72 M3
Division Artillery	93 M3
Engineer Battalion	15 M3
Div Trains HHC	2 M3
Ordnance Maintenance Battalion	4 M3
Medical Battalion	4 M3
Military Police Platoon	1 M3
Division total	451 M3
Tank Battalion, typical	
Tank Battalion (as of 15 Sept 43)	13 M3 Personnel, 3 M3 Ambulance
Medium, Special (as of 4 Dec 43)	9 M3 Personnel, 3 M3 Ambulance
Light Tank Battalion (as of 12 Nov 43)	12 M3 Personnel, 2 M3 Ambulance
Cavalry Division (as 1 August 1942)	
Reconnaissance Squadron	1 M2 & 2 M3
Engineer Squadron	7 M2
Cavalry Reconnaissance Squadron	
Part of Armored Div (as of 15 Sep 43)	32 M3A1 Personnel, 4 M3 Ambulance
Part of Armored Div (as of 9 July 43)	32 M3A1 Personnel, 4 M3 Ambulance
Nondivisional (as of 15 Sep 43)	26 M3A1 Personnel, 4 M3 Ambulance
Nondivisional (as of 9 July 43)	26 M3A1 Personnel, 4 M3 Ambulance
Tank Destroyer Battalion	
Tank Destroyer, Towed (as of 7 May 1943)	36 M3
Light and Medium Field Artillery Battalion	
Armored (as of 15 September 1943)	32 M3A1
Combat Engineer Battalion	
Armored 15 (as of Sept 43)	17 M2 or M3
Separate (as of April 42)	5 M2 or M3

Appendix C

Selected Scale Drawings

M3A1 Scout Car

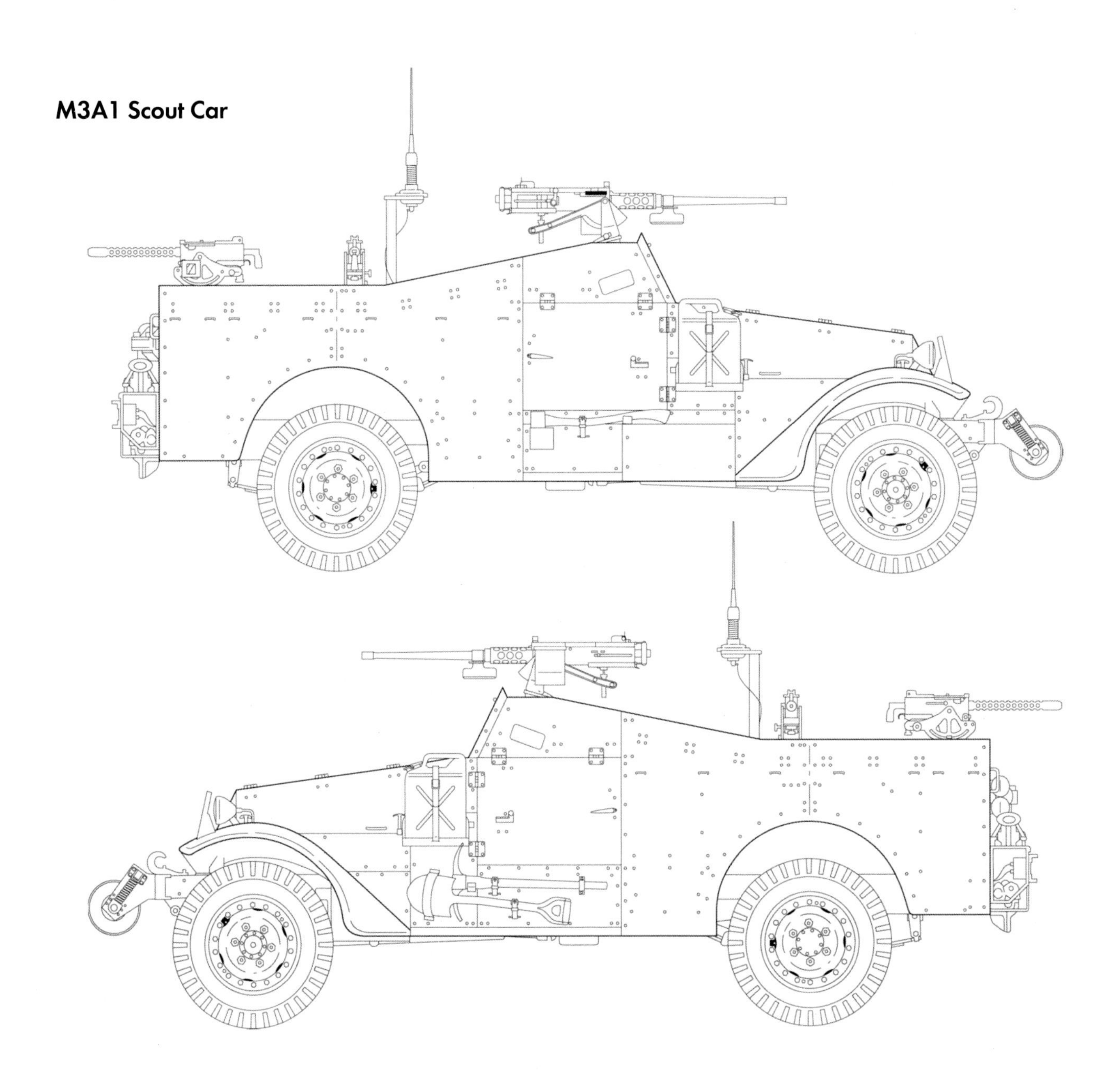

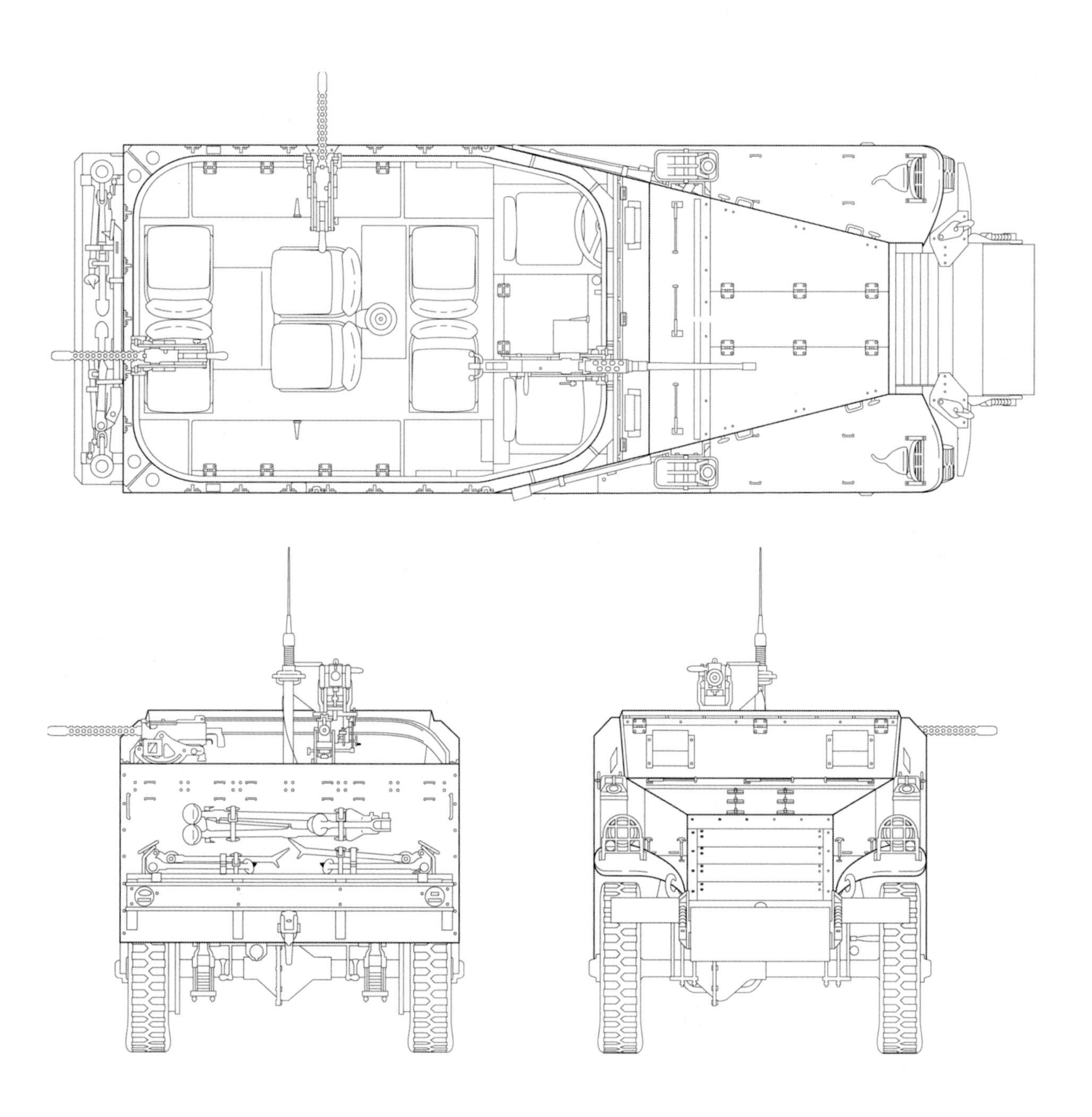

M2 Half-track Car

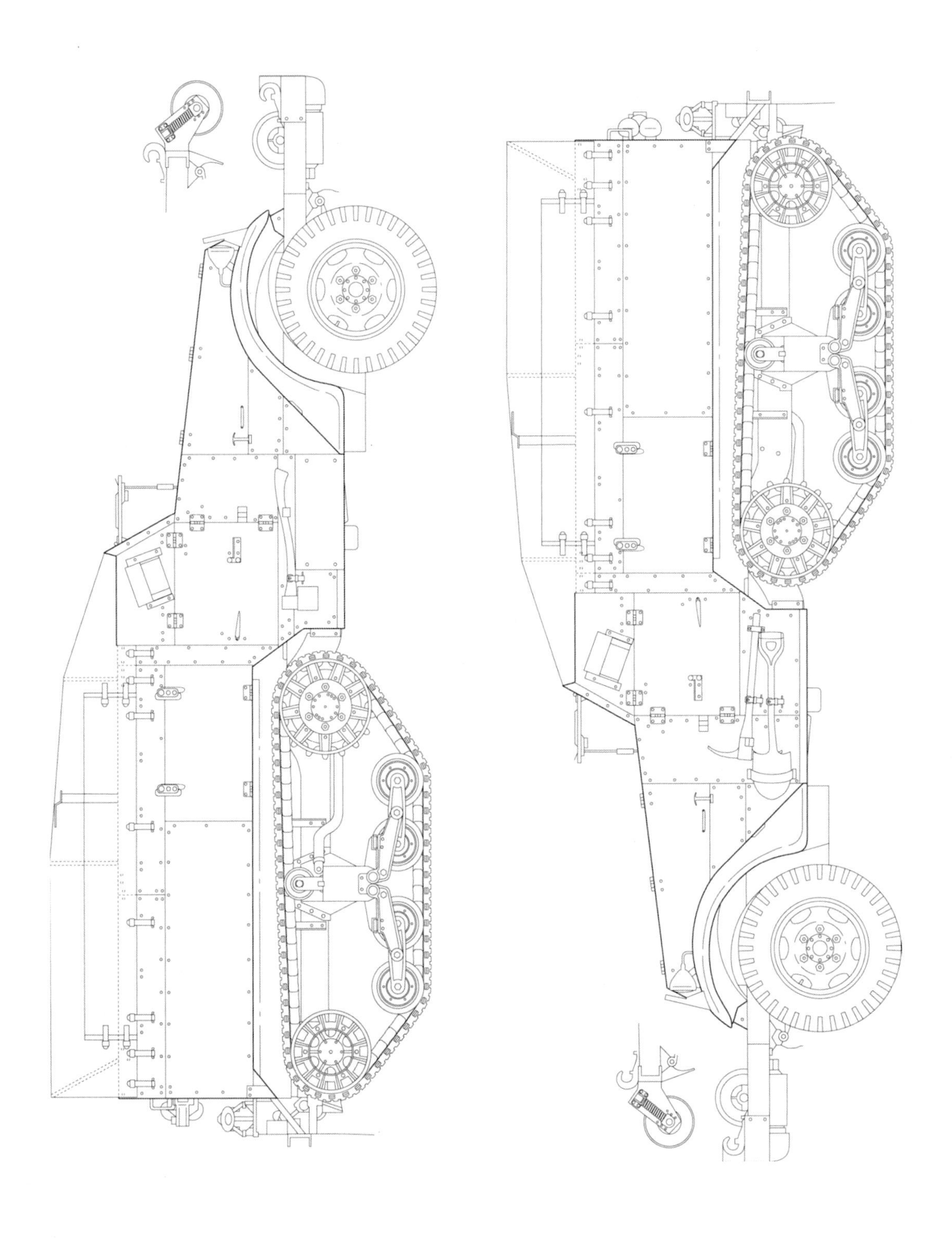

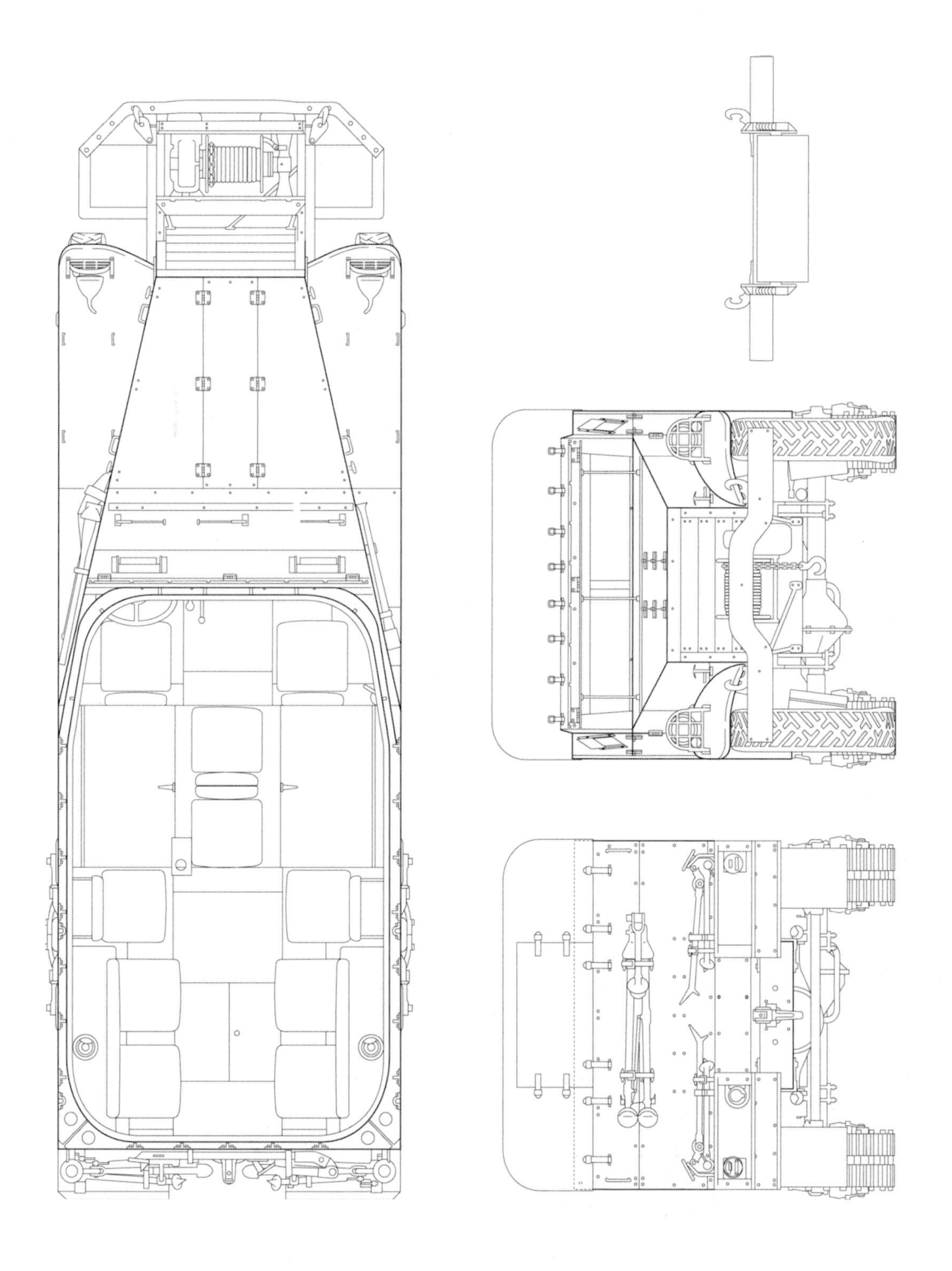

M3A1 Half-track Personnel Carrier

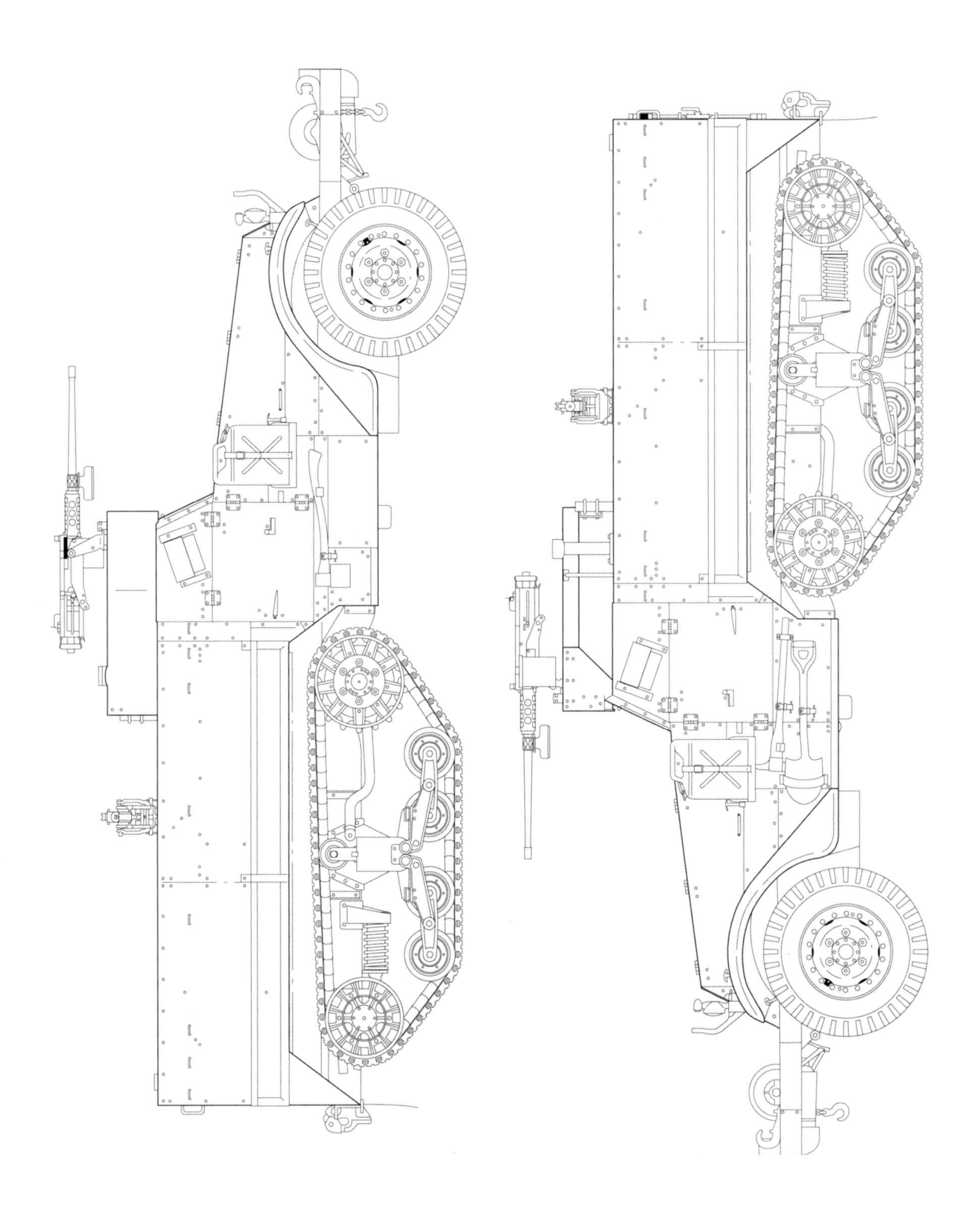

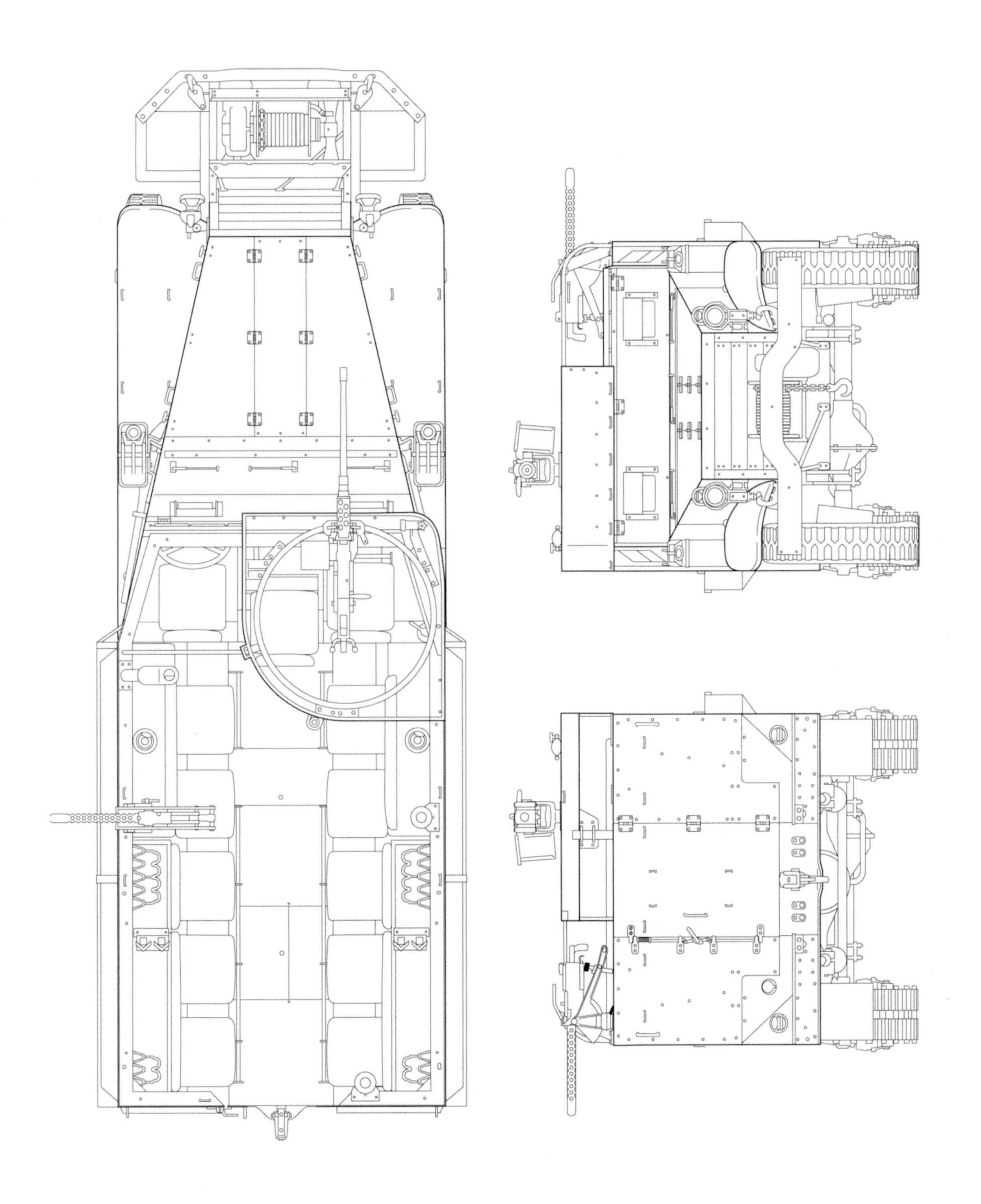

M5A1 Half-track Personnel Carrier

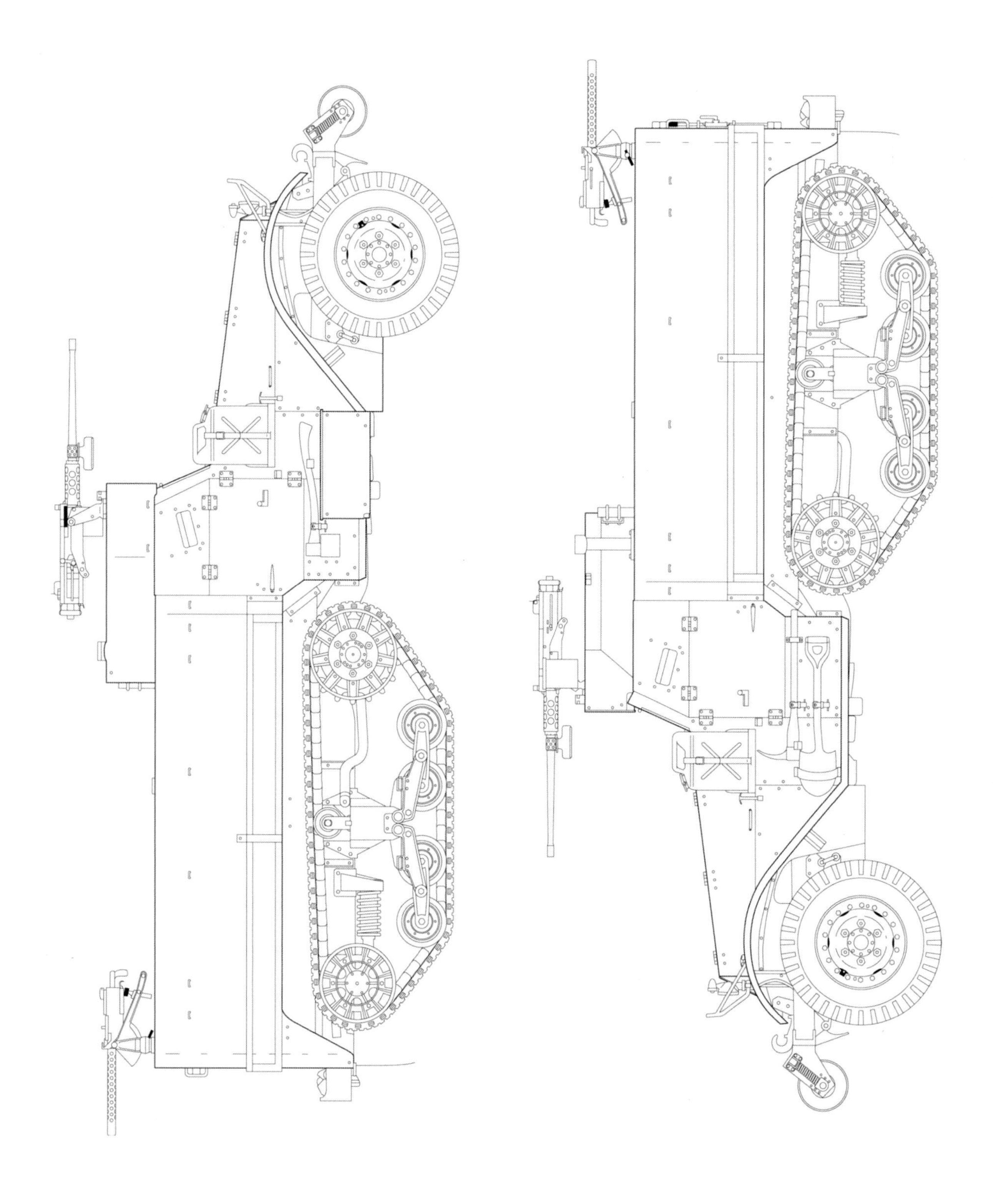

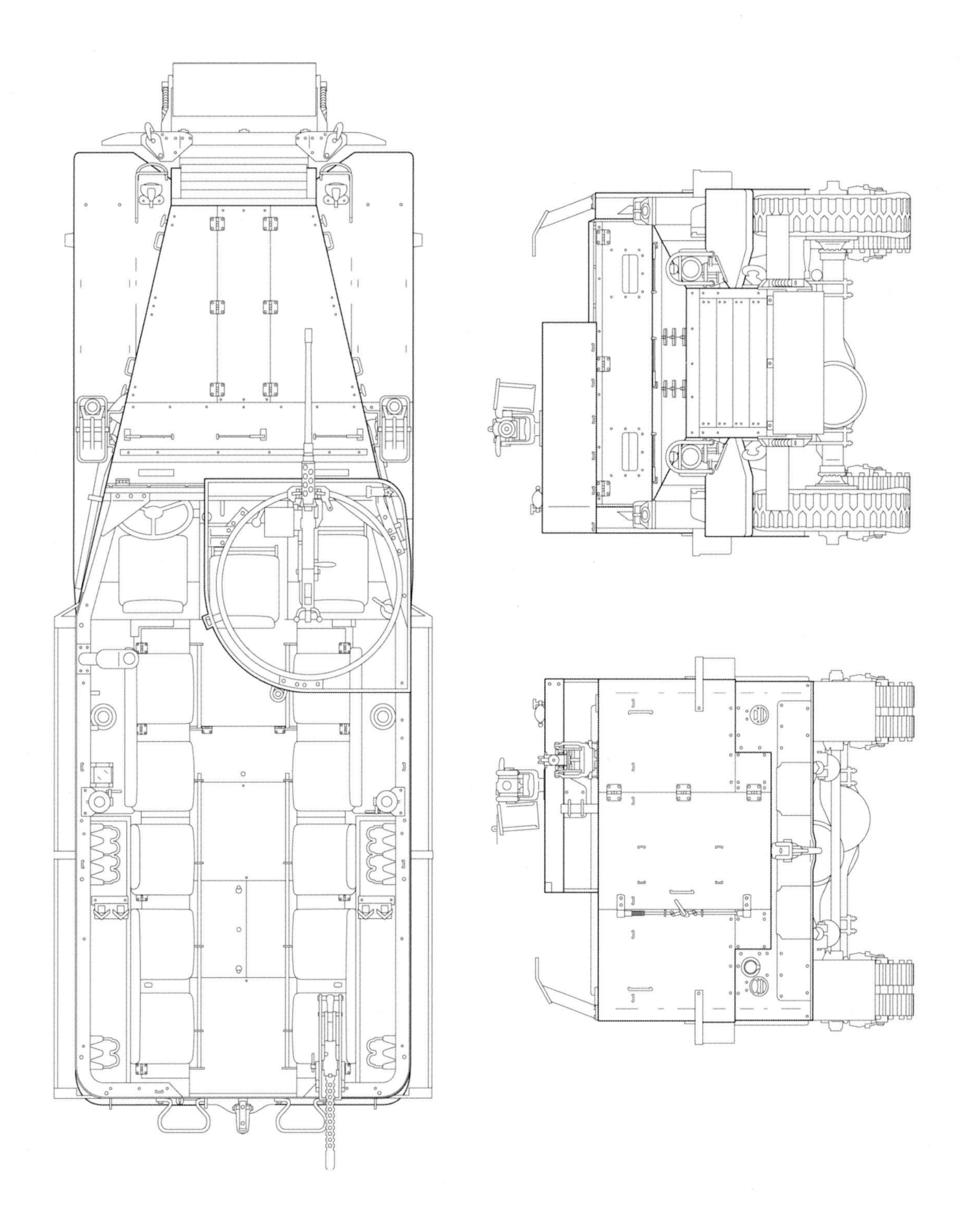

M9A1 Half-track Car

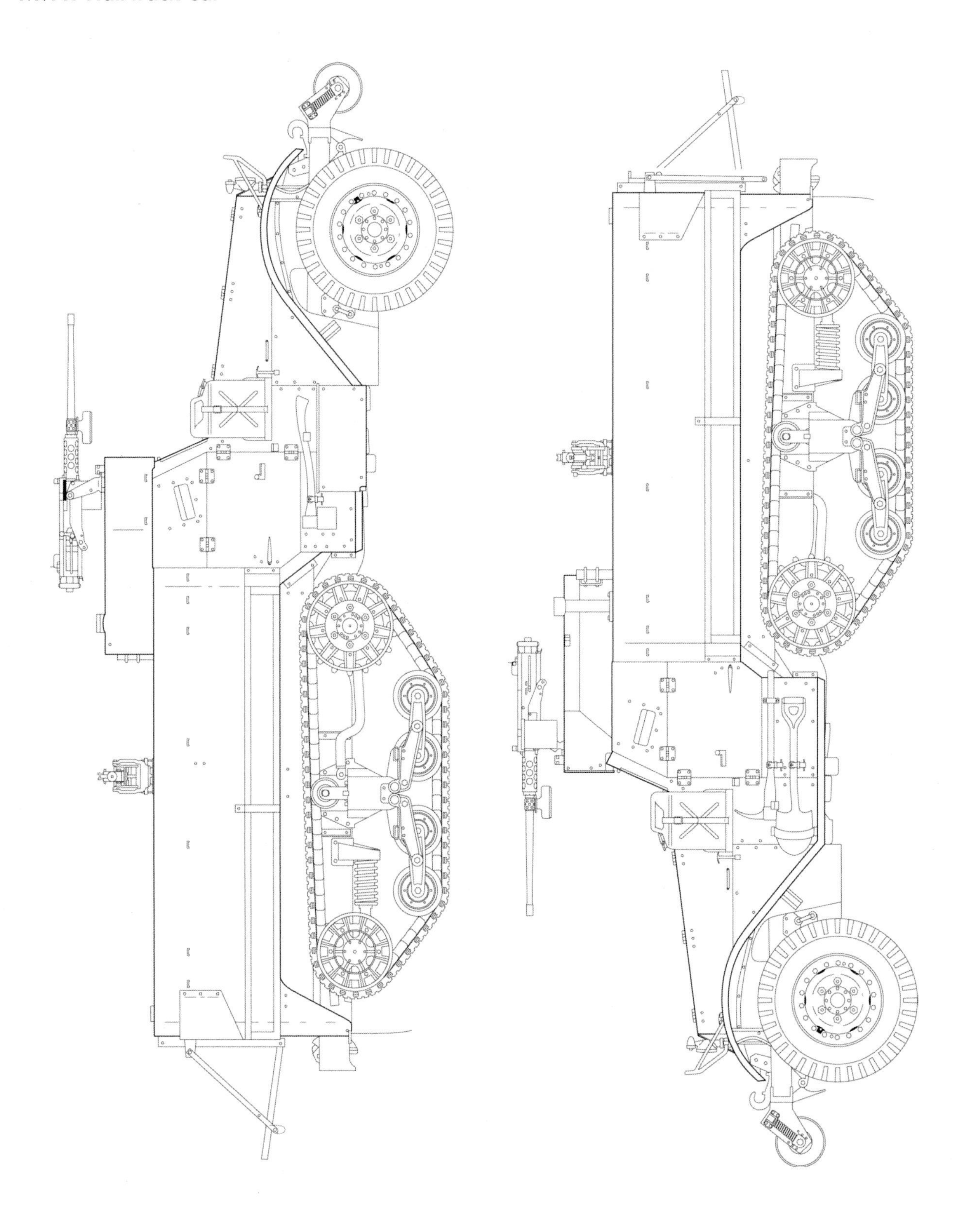

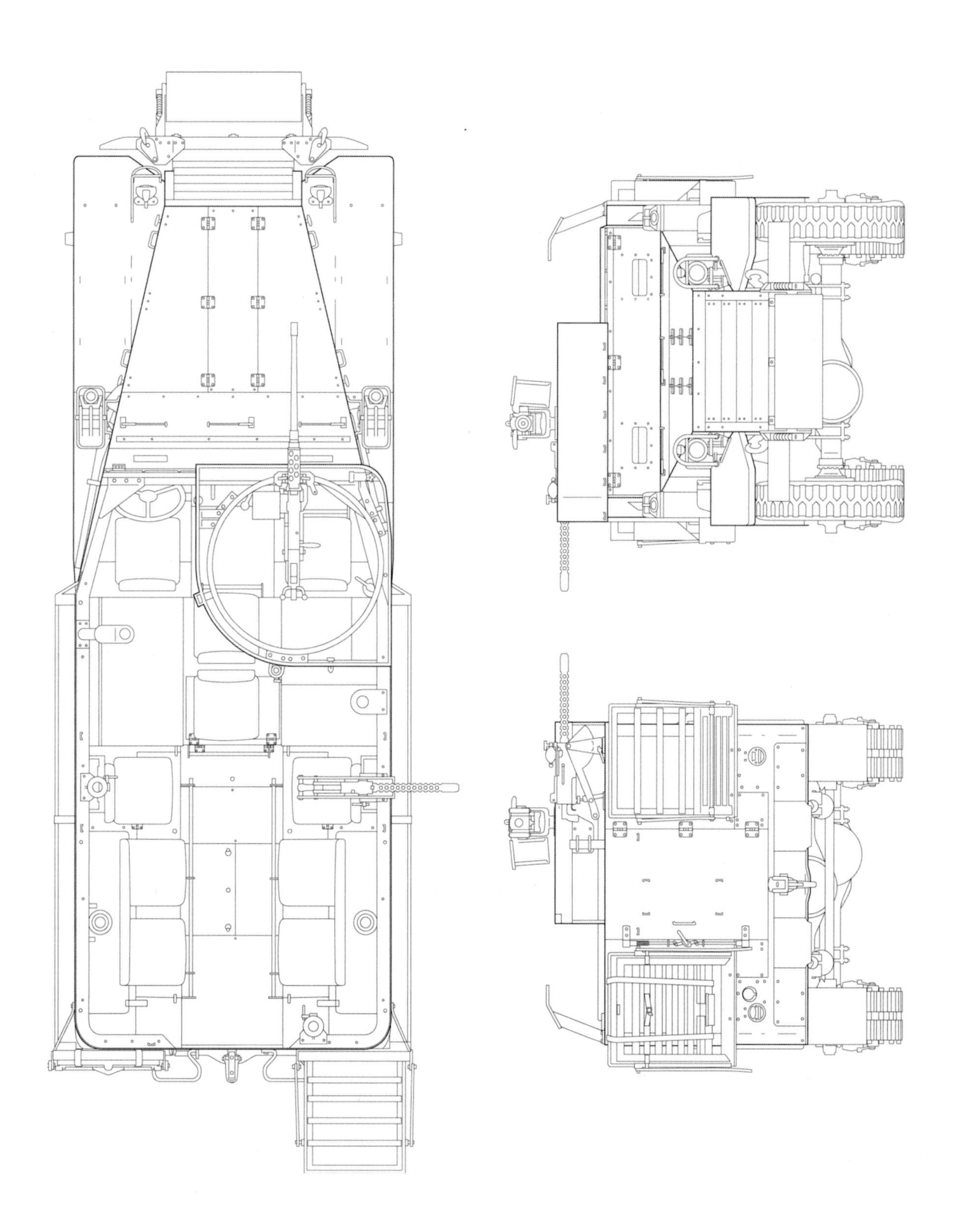

Bibliography

Monographs

Hunnicutt, R. P., *Half-track A History of American Semi-Tracked Vehicles*: Novato, CA: Presidio Press, 2001. ISBN: 0-89141-742-7

Stanton, Shelby L., *World War II Order of Battle*: New York, NY: Galahad Books, 1984. ISBN: 0-88365-775-9

Noville, G.O. & Associates, *Weapons Mounts for Secondary Armament*, Detroit, Mi: G.O. Noville & Associates, 1957.

Technical Manuals

ORD 9 SNL G-147; Half-track vehicles: Car, half-track, M9 (M2E5) (International Harvester Co.); carrier, personnel, half-track, M5 (M3E2) (International Harvester Co.); carriage, motor, multiple gun, M14 (International Harvester Co.); carriage, motor, multiple gun, M17. September 1, 1943

TM 9-707, Basic Half-Track Vehicles (IHC) (Personnel Carrier M5, Car M9A1, Multiple Gun Motor Carriage M14, and similar IHC Vehicles) May 21, 1943.

ORD 9 SNL G-102 List of All Parts for Half-Track Vehicles (White, Autocar and Diamond T), 1 December 1944

TM 9-1710C Ordnance Maintenance Chassis and Body for Half-Track Vehicles, September 11, 1942

TM 9-1711 Ordnance Maintenance White 160AX engine, Half-Track Vehicles, January 26, 1942.

TM 9-710 Basic Half-Track Vehicles (White, Autocar and Diamond T), 5 January 1942

TM 9-710 Basic Half-Track Vehicles (White, Autocar and Diamond T), 23 February 1944

TM 9-710 Basic Half-Track Vehicles (White, Autocar and Diamond T), 8 May 1953

ORD 7-8-9 SNL G-67 Ordnance Supply Catalog, Organizational Spare Parts and Equipment List Higher Echelon Spare Parts and Equipment List (Addendum) Spare Parts Catalog for Car, Scout, M3A1.

TM 9-705 Scout Car, M3A1 Operation and Maintenance, 19 February 1941

TM 9-705 Scout Car, M3A1 Operation and Maintenance, 26 October 1942

TM 9-1705 Scout Car, M3A1 Ordnance Maintenance Power Train (Axles, Propeller Shafts, Transfer Case, Transmission), 20 August 1942.

TM 9-1705C Scout Car, M3A1 Ordnance Maintenance Diesel Power Plant (Hercules DJXD Engine) 1 October 1942.

TM 9-1706 Scout Car, M3A1 Ordnance Maintenance Gasoline Power Plant (Hercules JXD Engine) 13 September 1942

TM 9-1709 Ordnance Maintenance Chassis and Body for Scout Cars M3A1, 22 September 1942

TM 9-1832A Hercules Engines, Series JX Models A, C, D, E-3, F, 17 May 1944

Documents

Project Supporting Paper 48 and 49 Half Tracks, Tractors and Carriers

Summary Report of Tank-Automotive Material Acceptances, September 1 1945

Statistical Work Sheets, To 1 August 1945, Volumes 41 and 42, Office of Chief of Ordnance – Detroit.

Official Munitions Production of the United States, by Months, July 1, 1940 – August 31, 1945; Civilian Production Administration, 1 May 1947.

Lend-Lease Shipments World War II, Office, Chief of Finance, War Department, 31 December 1946.

Alphabetic Listing of Major War Supply Contracts, Cumulative, June 1940 through September 1945, Civilian Production Administration, Industrial Statistics Division. 1946

Half-track Military Vehicles, A History and Continuing Survey, Edmund Lieberman, 1 October 1944.

Ordnance Department Armored, Tank and Combat, Vehicles 1940, 1941, 1942, 1943, 1944; Records Section, Statistics and Analysis Branch, Stock Control Division, Office Chief of Ordnance – Detroit, 15 December 1944.

The Armored Force Board, Fort Knox, Kentucky, Project Number 136-8, Stowage and Equipment List for 4 Litter Carrier (Evacuating) Half-Track, M3, 26 May 1943.

Numerous Ordnance Committee Minutes and Aberdeen Proving Ground test reports.

Above: *In an obviously posed photo, a G.I. wearing an M1917 helmet aims an M1 Garand rifle through the right side window of an M3 half-track. Details of the lowered window shield are visible, including the vision port and several footman loops.* **Below:** *A soldier assumes a defensive position alongside an M3 half-track at Fort Knox, Kentucky, in June 1942. The styles of headlight and brush guard, as well as the front wheel, are characteristic of early-production M2 and M3 half-tracks. (LOC, both)*

Members of a half-track crew check their equipment, including a .50-caliber machine gun, in England prior to the D-day invasion of Normandy. From left to right, they are Cpl. John Hartlage, Capt. Edward L. Smith, and Pvt. George Roberts. The flexible hose to the far left appears to have been the air intake of a deep-water fording kit. (NARA)

Above: *The crew of an M2 half-track of a tank-destroyer unit conducts an antiaircraft watch at a desert training base in the United States in 1944. The vehicle was well camouflaged with brush on the front and netting on the rear.* **Below:** *An officer in a very dusty M2 half-track points the way forward during 7th Armored Division maneuvers in the Southwest during World War II. The vehicle has the early-type headlights and a mine rack to the rear of the stowage-compartment door. (NARA, both)*

Above: *A restored M2A1 half-track is on display at a War and Peace military vehicle show in the U.K. The door of the stowage compartment is open, showing the machine-gun ammunition boxes stacked inside the compartment.* **Below:** *An M3A1 half-track personnel carrier at the War and Peace show is fitted with a tarpaulin with a panel open to allow the side-mounted .30-caliber machine gun to protrude. Crates, tarpaulins, ammo boxes and camouflage netting are secured to the rear racks. (PAS, both)*

GANGWAY for VICTORY!
Slashing, tank-busting Half-Tracks by Autocar . . . Combination Gun Motor Carriages, deadly against strafing aircraft . . . swift, armored mobility of many types for the Army, Navy, Marine Corps, and Air Forces. At every invasion point of American forces, Autocar has helped to win the beachhead. . . . Victory is a great teacher. When the war is won, this heavy-duty fighting will be transformed into heavy-duty, low-cost-per-mile transportation—these heavy-duty fighters into the heavy-duty trucks that bear a famous name.
AUTOCAR TRUCKS
MANUFACTURED IN ARDMORE, PA.
SERVICED BY FACTORY BRANCHES FROM COAST TO COAST
KEEP PRICES DOWN!
Buy and
Hold War Bonds

DIAMOND T ON DUTY

...in defense of Freedom!

BEFORE the finest military vehicles in history could roll by thousands from Diamond T's expanded production lines, there had to be a Diamond T organization that, through years of experience, knew how to design and build Super-Service Motor Trucks.

Today, Diamond T is building versatile armored "half-tracks" such as the one pictured above, as well as great six-wheel-drive, four-ton "prime movers" and special Diesel powered vehicles of enormous capacity for carrying army tanks.

Hauling the nation's freight loads may not be so spectacular as, say, picking up and lugging a disabled army tank, towing artillery, or carrying men and equipment in armored vehicles at high speed. *But it takes the same kind of truck stamina.*

Every one of the Diamond T Super-Service Trucks which we are able to build in restricted numbers for vital civilian service, becomes at once a mighty force in conserving America's man-hours, fuel, oil and materials.

Diamond T Motor Car Company, Chicago

Commercial models in all sizes and types to meet the needs of defense transportation.

DIAMOND T

★ *Trucks* ★

Trucks That Rain Death Upward

THE INTERNATIONAL HALF-TRACK is a truck that carries its own pavement. It can speed over bog, sand, mud and mountain . . . carrying armed-to-the-teeth personnel to seize and hold a position, or toting fast anti-aircraft firepower that rains death upward.

The International Half-Track is proving on the world's battlefronts that it can take it, as well as dish it out. It should. It's a brother under the armor to the International Truck that was the *largest selling heavy-duty truck* on the market when civilian trucks were still being made.

When the story of this war is written, trucks will contribute one of the most glorious chapters. A vital part of this war is being waged on the highways of America, where trucks haul materials to keep the wheels of America's war production turning, and other trucks haul food supplies to feed America's great army of industry.

Trucks must work harder and longer, to the last possible mile, because there aren't any new trucks to take their places. That means that every truck on the road today must be babied and serviced to give better and longer wear than was ever expected of trucks before.

And International civilian truck service—*the largest company-owned truck service organization in the world*—is now a wartime truck *service*. More alert and more efficient than ever to keep your trucks on their jobs. Whether they're International Trucks or any other make, bring them to an International Branch or Dealer. You'll find International service close at hand—pledged to keep your trucks rolling—*pledged to Victory!*

INTERNATIONAL HARVESTER COMPANY
180 North Michigan Avenue Chicago, Illinois

MAJOR WAR PRODUCTS BUILT BY INTERNATIONAL HARVESTER

Half-Track Military Vehicles Torpedoes
Artillery Prime Movers Automatic Airplane Cannon
Oerlikon Gun Mounts Military Trucks
Military Tractors Steel Products for Military Use
Aerodrome Control Trucks Armored Scout Car Hulls
High Speed 155 mm. Gun Carriages Gun Loaders
Airplane Engine Cowling Assemblies
Tank Transmissions Blood Bank Refrigerators
Shells Gun Carriages Adapter Boosters Trackers
Marine Corps Invasion Ice Chests

Four Harvester Plants have been awarded the Army-Navy "E" for Excellence.

★ **INTERNATIONAL TRUCKS**

ANOTHER EXAMPLE OF WHAT A BETTER TRUCK CAN DO

Trucks You Don't Steer...You Aim

More than 2500 armored White Scout Cars, of the four-wheel type pictured above, have been delivered to the army five months ahead of schedule.

In action, the White Scout Cars carry a crew of eight men. The armament consists of two 30 caliber machine guns and one of 50 caliber.

The shortest distance between two points is a straight line, but roads seldom run that way. To the mechanized cavalry of the U. S. Army, this meant just one thing . . . design and build armored vehicles that can leave the traveled road almost anywhere enroute and not be A. W. O. L. when the engagement begins. A dry river bed or a route through a stone quarry could be preferable to U. S. Route 1. The result: the Government's order for more than 2500 White Scout Cars has been completed five months ahead of schedule. White Super Power Half-Tracs . . . trucks that you *can* steer and easily, but if necessary you can *aim* over the roughest terrain men can ride over . . . are rolling off the production line and soon 5,000 will be added to the Army's mechanized forces. As in all commercial fields, White quality makes it possible to rely on transportation equipment to do unusual jobs unusually well.

THE WHITE MOTOR COMPANY, Cleveland

Builders of the complete line of White Super Power Trucks, city and inter-city coaches, Safety School Busses and the famous White Horse.

FOR 40 YEARS THE GREATEST NAME IN TRUCKS